PROTECTED AREAS, SUSTAINABLE LAND?

T0231678

Protected Areas, Sustainable Land?

Edited by

CATHERINE AUBERTIN and ESTIENNE RODARY
Institut de recherche pour le développement, France

Translated from the French by Laurent Chauvet, with the collaboration of
Chris Jackson and Teresa Dirsuweit

LONDON AND NEW YORK

First published 2011 by Ashgate Publishing

2 Park Square, Milton Park, Abingdon, Oxon OX14 4RN
711 Third Avenue, New York, NY 10017, USA

Routledge is an imprint of the Taylor & Francis Group, an informa business

First issued in paperback 2016

British Library Cataloguing in Publication Data
Protected areas, sustainable land?.
   1. Protected areas. 2. Sustainable development. 3. Environmental management.
   I. Aubertin, Catherine. II. Rodary, Estienne.
   333.7'2-dc22

Library of Congress Cataloging-in-Publication Data
Protected areas, sustainable land?/[edited] by Catherine Aubertin and Estienne Rodary.
   p. cm.
   Includes index.
   ISBN 978-1-4094-1235-9 (hardback : alk. paper)
   1. Protected areas—Developing countries—Case studies. 2. Sustainable development—Developing countries—Case studies. 3. Nature conservation—Government policy—Case studies. I. Aubertin, Catherine. II. Rodary, Estienne.
   S944.5.P78P76 2010
   333.72—dc22

                                  2010049364

ISBN 978-1-4094-1235-9 (hbk)
ISBN 978-1-138-25097-0 (pbk)

Contents

PART III   NEW CONSERVATION TERRITORIES

List of Plates

The Plates are located between pages 110–111.

Note: Plates 4 to 10 are also available on http://www>carto-conservation.net.

List of Figures

List of Tables and Boxes

Tables

Boxes

Notes on Contributors

Bruce Albert, anthropologist, is professorial fellow at the IRD, Paris, and associate researcher at the Instituto Socioambiental, São Paulo. Bruce Albert has been working with the Yanomami in Brazil for thirty-five years. These past years, he has published four books in Portuguese and French: *Pacificando o Branco. Cosmologias do contato no Norte amazônico* (2002), *Yanomami. L'esprit de la forêt* (2003), *Urihi a. A terre-floresta Yanomami* (2009) and *La chute du ciel. Paroles d'un chaman yanomami* (2010) co-authored with Davi Kopenawa, shaman and spokesperson for the Brazilian Yanomami.

Fano Andriamahefazafy, economist, is researcher at the Center of Economy and Ethics for the Environment and the Development – Madagascar (C3EDM) at the Department of Economics, University of Antananarivo. He obtained his Ph.D. in economics from the University of Versailles Saint-Quentin-en-Yvelines (France). His research interest includes applied economics of development policies, peasant economy, and payment for environmental services.

Catherine Aubertin is an economist, professorial fellow at the IRD. She has been coordinating several research programs on economic instruments applied to development and environment policies. Her main fields of research include international conventions, biodiversity and ecosystem services markets, protected areas, sustainable development with a North-South approach. She is member of the editorial board of the journal *Natures, Sciences, Sociétés*.

Marie Bonnin is senior research fellow at the IRD. She specialised in international environmental law. Her current research interests include biodiversity, protected area, ecological network, ecosystem services, tourism and spatial maritime planning. She published *Les corridors écologiques: Vers un troisième temps du droit de la conservation de la nature?* in 2008.

Jean Boutrais is a geographer, professorial fellow at the IRD, Paris. He has worked mainly in Cameroon but also in Centrafrique, Niger and Burkina-Faso. He is a specialist on Fulani pastoralism, about which he has written extensively. He is the author of *Des Peul en savanes humides: développement pastoral dans l'Ouest centrafricain* (1988), *Hautes Terres d'élevage au Cameroun* (1995–6) and co-editor of *Figures peules* (1999), *Patrimonialiser la nature tropicale* (2002) and *Patrimoines naturels au Sud* (2005). He is the scientific editor of *A travers champs*, a book series published by the IRD.

Ambroise Brenier has conducted a Ph.D. in Madagascar and French Polynesia on participatory monitoring of reef fisheries. More recently, he worked as a monitoring and evaluation officer with the International Foundation of the Banc d'Arguin, a marine conservation NGO supporting marine protected areas in West Africa. He is now marine technical director at the Wildlife Conservation Society in Madagascar.

Stéphanie M. Carrière is senior research fellow in ethnoecology at the IRD. Her research focuses on traditional farming practices and their effects on tropical forests dynamics, as well as on the role of agroecosystems in biodiversity conservation, mostly in Central Africa and Madagascar. She is also interested in traditional knowledge, local resources management and the relevance of ecological concepts and data in the construction of biodiversity conservation policies in developing countries. She has published articles and books on the subject, including *Les orphelins de la forêt* (2003).

Christian Chaboud is economist, professorial fellow at the IRD in the Mediterranean and Tropical Fisheries Research Centre in Sète (France). He was coordinator of the program "Economic and social stakes of biodiversity in a context of great poverty: the south-west region of Madagascar" funded by the French Institute on Biodiversity, and was in charge of the economic component of the National Shrimp Research Program in Madagascar. He is the editor of several books on natural resource management: *Du bon usage des ressources renouvelables* (2000), *La ruée vers l'or rose* (2002), *Madagascar face aux enjeux du développement durable* (2007).

Gilbert David is senior research fellow in marine and island geography at the IRD. Based in Brest, he has led several research programs dealing with marine protected areas, including relationships with watersheds, in Reunion Island (2002–2006 and 2004–2005), Madagascar (2005–2008) and Pacific islands (2006–2010).

Pascale de Robert is senior research fellow in anthropology at the IRD. She has worked on social change, relationships to 'nature' and territory in Andean and Amazonian communities. Currently based in the Museu Goeldi in Brazil (MPEG, Belém), she is involved in research programs on agricultural biodiversity and sustainable development among the Kayapo community.

Jocelyne Ferraris is professorial fellow at the IRD, specialised in ecological data analysis. She has been head of the research unit on "Reef Communities and Uses of Indo-Pacific Coral Ecosystems". Her current research concerns interactions between living marine resources, their environment and uses, in particular fishing. She is particularly interested in actors' strategies, impact of human practices on resources and indicators of natural and anthropogenic perturbations, such as marine protected areas.

Geoffroy Filoche is a lawyer and senior research fellow at the IRD. He works on biodiversity law, intellectual property standards and multiculturalism policies directed at indigenous and local peoples in South America.

Géraldine Froger is associate professor in economics at the University of Versailles Saint-Quentin-en-Yvelines. She is a researcher at the CEMOTEV (Centre for the Study on Globalisation, Conflicts, Territories and Vulnerabilities) with wide experience in the ecological economics field. Her papers have been published in journals such as *Ecological Economics, International Journal of Sustainable Development, Mondes en Développement, Développement Durable et Territoires*. She has been involved in several European and French research projects on sustainable development. Her research focuses on environmental valuation and decision-support tools, governance and environmental policies in developing countries, tourism and sustainable development.

Florence Galletti is lawyer, senior research fellow at the IRD. She works in the Mediterranean and Tropical Fisheries Research Centre in Sète (France) with a focus on governance and law of natural resources in the Indian Ocean. She previously worked in Madagascar where she was responsible for the programme "Governance, Environment and Sustainable Development in Interdisciplinary fields in Madagascar". Her work, books and features cover areas on comparative law, development law, environmental law, law of the sea and marine or coastal resources.

Dominique Hervé is professorial fellow at the IRD, Madagascar. With a Ph.D in agronomy (National Agronomic Institute of Paris-Grignon), he is specialised in cropping and farming systems and environmental modelling. He spent 15 years leading multidisciplinary research teams in the South-American Andes pairing social scientists and ecologists on land-use dynamics issues and studying how fallow operates in long-grazed fallow cropping systems. Since 2002, he has been working on forest transition to agriculture in Madagascar, bringing together environmental scientists (agronomists, ecologists, geographers) and modelers (computer scientists and mathematicians). The products of his research are prediction and control tools for planners to trade-off rainforest conservation and agricultural development.

Anne Élisabeth Laques is professorial fellow in geography at the IRD, Montpellier. Using remote sensing, she works on biodiversity indicators and landscape dynamic at local scale in order to inform public policy.

François-Michel Le Tourneau is senior research fellow at the French National Center for Scientific Research (CNRS). He was associated professor in the Centre for Sustainable Development in Brasilia during five years (2002–2005 and 2008–2010). As a geographer, his research focuses on the human dynamics of the Brazilian Amazon, with a special emphasis on isolated areas and/or areas devoted to "traditional populations". He has written extensively on the Yanomami Indians, and is currently responsible for a research project combining geographical and anthropological approaches for studying the "territoriality" of traditional communities.

Philippe Méral, economist, is a senior research fellow at the IRD, Montpellier. Specialised in economics of conservation in developing countries, he currently coordinates a research program on ecosystem services in France, Costa Rica and Madagascar. He previously led a research program on environmental policy and local management of the environment in Madagascar (2001–2005).

Johan Milian is associate professor at the University of Paris 8 Saint-Denis. Geographer, he is specialised in human ecology and spatial development, in particular in relation to conservation policies and management. His researches concern mountain rural areas in France, Spain and Morocco.

Florence Pinton is professor of sociology at AgroParisTech and research associate at the IRD. Her works deal with the way collective action operates in environmental public policies, territorial construction and the developement of environmental standards. She is editor of *La construction du réseau Natura 2000 en France* (2007). She was also editor of *Les marchés de la biodiversité* (with C. Aubertin and V. Boisvert, 2007).

Ando Rabearisoa (MSc in Economics) is the marine program coordinator at Conservation International Madagascar. Since 2006, she has collaborated with scientists from James Cook University in Australia to conduct researches on of marine resources governance in Madagascar, the results of which were published in *Marine Policy Journal.* In November 2009, she earned the overall best student presentation during the Western Indian Ocean Marine Science Association's Symposium in La Réunion.

Hervé Rakoto Ramiarantsoa is geographer, professor at the University of Poitiers, and research associate at the IRD, Orléans. He has studied the relationships between societies and their environment in Madagascar, especially using landscape studies (1995, 2000). He has also worked on community-based management of forest resources and its impacts on biodiversity conservation in the highlands of Madagascar (2004–2009).

Estienne Rodary is senior research fellow at the IRD. With a background in geography and political science, he is specialised in biodiversity conservation policies, protected areas and cosmopolitanism. He is currently visiting researcher in the Development Studies Programme, University of the Witwatersrand (Johannesburg, South Africa), where he works on transnational parks and the politics of environmental connectivity. He has published in several journals such as *L'Espace Géographique, Revue Tiers Monde, Conservation and Biodiversity, Mondes en Développement,* and is co-editor of *Conservation et développment. L'intégration impossible ?* (2003). He is editor in chief of the journal *Écologie & Politique* and member of the editorial board of the journal *L'Espace Politique.*

Introduction
Sustainable Development, A New Age for Conservation?

Catherine Aubertin, Florence Pinton and Estienne Rodary

Since the genesis of sustainable development discourse, the place of protected areas in environmental policies has fuelled heated debates. As the main subject of conservation policies throughout the 20th century (while nature protection remained a marginal issue consigned to peripheral reserves), protected areas have become a central issue which has extended to the management of the global environment. The aim of this publication is to assess whether the environment has emerged as a primary referent for public policies, or, on the contrary, it remains secondary to the imperatives of economic development and resource exploitation. Applied to protected areas, the debate amounts to addressing whether they should become the tool of sustainable development policies, or they should be confined to the more restricted role of protecting 'outstanding' biodiversity. In this publication these issues will be addressed through an evaluation of the extent to which protected areas become zones of experimentation for sustainable development.

Since nature conservation policies are at the core of the notion of sustainable development, it comes as no surprise that such policies share the ambiguities of sustainable development. During the last decades, this notion led to important advances in environmental policies: from participative projects during the 1980s to regional and global approaches initiated during the 1990s and, more recently, to the 'back to the barriers' movement that tried to return to stricter forms of conservation.

These transformations were the result of political choices: they reflected empirical findings as well as the evolving representations of conservation actors. This had two effects. First, conservation policies became more complex as policy goals and actor categories were restructured. Indeed, during the last 30 years, the historical sector of conservation has opened up towards society, with a restructuring of the classic forms of the 'field' of conservation, where new actors and new power relations are challenging the previous order. Second, it also led to interactions between the local and the global, which in turn caused important redefinitions in the methods and tools used in conservation policies. Nowadays, national public policies are required to integrate international or transnational factors. These either partly invalidate the classic approaches to conservation, or shift the focus towards other dimensions linked to it, for instance the ecosystem connectivity, the importance of flows, liaisons and networks, or the influence of global warming.

Because it is now recognised that protected areas can support market activities at the local level, such as the craft industry or ecotourism, but also at the global level with ecosystem services linked to biodiversity protection and carbon storage, and because conservation policies can concern spaces outside of protected areas, an extreme scenario could be the disappearance of protected areas altogether. This theoretical scenario must be the context in which one discusses the objectives of generalising sustainable development practices, or mainstreaming biodiversity conservation in all sectors of society and not only in protected areas. Thus, the questions facing conservation policies no longer concern only the experts of 'nature' protection who no longer have adequate legitimacy to act alone, in defining and applying conservation practices.

Within this frame of reference, this introduction reviews the main orientations of conservation: the invention of parks and their consolidation within the state, the community-based policies and, more recently, the attempts to return to a stricter form of conservation. Indeed, the issues presented in all the chapters of this publication are underpinned by the debates and orientations that have been shaping the world of conservation for over a century.

The Main Orientations of Conservation

The success of protected areas is such, today, that they can be considered as one of the main land tenure system on the planet, occupying 12% of the land and an exponentially increasing amount of the marine surface area. Yet, nature conservation and its techniques are a fairly recent Western invention. The history of conservation policies is linked to environmental representations of different social groups whose spatial appropriation strategies have, more often than not, been conflictual (Adams 2004).

The Invention of the Parks   Modern nature protection began during the 19th century in North America and Europe (and its colonies), and was based on two great institutional traditions of nature protection: the associations for the protection of nature, and the forestry administrations.

In North America, the descendants of settlers sought to protect what remained of the wilderness, mainly forests. The first parks appeared with the creation of the Yellowstone Park in 1872, to preserve the wilderness and the landscape as they first appeared to White Americans. The encounters of Europeans with the tropics also resulted in policies based on protectionist concerns (Grove 1995). In many British overseas territories (e.g. Canada, Australia, New Zealand, India and South Africa), protected areas were already created by the end of the 19th century (Rodary and Milian, this publication). In reality, when the first settlers arrived in America, or in the tropics, the wilderness they were coveting was obviously not uninhabited. However the myth of the untouched wilderness constituted one of the strategies for the appropriation of space, and for the control of indigenous populations.

At the end of the 19th century, the protectionist movement fragmented into the supporters of preservation and the proponents of conservation (Nash 1967; Larrère and Larrère 1997). Whereas the former campaigned for the creation of sanctuaries for remarkable natural areas, the latter called for reserves that could secure 'sustainable' exploitation of natural resources. In any case, these two trends found themselves marginalised by the capitalist dynamics of natural resource exploitation. This marginalisation created a 'sector' of conservation policy that intervened only in reserves and parks: the more the environment outside those protected areas was transformed, the more conservation policy focused on them.

In the old rural civilisations of Europe and the Mediterranean Basin, the protection of the wilderness could only concern small areas scattered within countries that were largely transformed by humans. The policies have focused on protecting endangered species, exceptional natural environments and picturesque landscapes. At the beginning of the 20th century, foresters and tourism associations began collaborating with naturalists to promote nature protection and to contribute to the advent of national policies in this domain (Selmi 2006). In this regard, the conservation task assigned to the state served the public interest through environmental restoration, nationalisation and profitability (Viard 1990).

The park concept was also adopted by relatively vast countries with areas considered to be almost undisturbed by human activities. This is the case for Brazil where, from the 1930s, conservationist policies were tentatively developed, to save what was considered to be a national heritage. The first protected areas were created in the urbanised South of the country, and subsequently followed the expansionist movement of the society (Barreto Filho 2004). Closely related to the advance of pioneering approaches, they embraced the model of integral protection, thereby excluding local populations. This led to an increase in the conflicts between local populations and protected areas worldwide, thereby challenging the legitimacy of these procedures.

In France, where the creation of national parks turned out to be politically difficult, the first parks were established in the colonies (e.g. Indochina, West and Equatorial French Africa, and Madagascar). From 1925 onwards, these colonies became places of nature protection experimentation (Berdoulay and Soubeyran 2000; Ford 2004). In metropolitan France, the first regulations relating to eminent domain were voted on in 1958, and the first national park, the Vanoise National Park, was created in 1961. The park was conceived as a core area, protected from human activities and isolated from the outside world by peripheral zones, as such following the preservationist model. In time, this model of protected areas changed with a series of crises experienced by rural society: the modernisation of agriculture and industrial development, the impetus given to land use planning and, finally, the mobilisation of scientists joining associations of nature protection. Finally, the links between protected areas and agricultural, pastoral or even forestry practices were acknowledged: this led to the invention of regional nature parks in 1967, as a break from the national park model. Since then the regional nature park has become a land use planning and protection tool over which elected officials have control.

During that period, in terms of international politics, environmental concerns drew closer to development ideology. This in turn transformed the main orientations of the global movement of nature conservation.

The Turn of Participative Management   The 1970s have indeed been a period of profound crises for conservation policies. The management of biodiversity by the state led in many cases to serious social and ecological crises, and to strong criticism vis-à-vis centralised and state-controlled nature conservation. *The World Conservation Strategy*, published in 1980 by the International Union for Conservation of Nature (IUCN), the World Wildlife Fund (today the World Wide Fund for Nature – WWF) and the United Nations Environment Programme (IUCN et al. 1980), tried to address these criticisms and was the first international document to use the term 'sustainable development' (Vivien 2005). The initiative from UNESCO to implement the biosphere reserves within the framework of the Man and the Biosphere Programme, and the first experiences of community-based management of nature, renewed all the objectives and operating modes of conservation. The management of resources was entrusted to local actors, by presenting community identities and private economic issues as central for conservation policies. This evolution characterised a radical change in the perception of relations between human activities and the permanence of nature. At the beginning of the 1990s, the social sciences participated in this movement by acknowledging the role played by traditional knowledge within local ecosystems. In parallel, the Convention on Biological Diversity legitimated the necessity for a profound overhaul of protected area models, by promoting management at the ecosystem level, instead of at the level of the species. This was the last step of a larger movement for which the goal was no longer to freeze nature into sanctuaries, but to protect the evolutionary potentiality of ecological processes, while maintaining certain human practices enabling the populations to benefit from their conservation efforts. Protected areas were to be integrated into territories occupied and laid out by societies. Conservation programmes moving in this direction were implemented as early as the 1980s, through 'participation policies', 'community-based conservation', 'sustainable development and use', and 'integrated programmes for development and conservation' among others. In 1996, the World Commission on Protected Areas of the IUCN (IUCN-WCPA) and the WWF produced a document entitled *Principles and Guidelines on Indigenous and Traditional Peoples and Protected Areas* (IUCN and WWF 1999), which highlighted the need to manage these areas together with indigenous people and to respect their knowledge about the environment.

The participation of the local population, which is political (through decentralisation), and at the same time economic (through the redistribution of revenues generated by the natural resources and through local employment), became dominant in the conservation rhetoric. The will to bring together conservation and development provoked a very strong trend, reinforced by the concept of biodiversity associating ecological diversity with the diversity of human practices, and bringing together scientific analysis and political action. The collaborations between various organisations multiplied to give rise to an important global epistemic community producing both global expertise and advocacy on conservation issues. For conservation NGOs, this was a political opportunity to become aligned with global governance and to become the central operators of sustainable development (Dumoulin and Rodary 2005). For the conservationists, if environment were to become the heart of all sectors of public action, 'integrated' conservation policies could be extended to many sectors that were historically unrelated to nature protection, such as land use planning or agriculture.

Back to the Barriers   Local participation naturally creates expectations that have often ended in disappointment. In this regard, various criticisms have questioned community-based approaches. A first set of critics have focused on a political analysis and have showed that conservation based on local mobilisation is often a failure (McShane and Wells 2004; Brosius et al. 2005; Spiteri and Nepal 2006; Shackleton et al. 2010a; 2010b). This is due for instance to the fact that egalitarian participation in local socio-political structures was an impossibility, due to overriding representative systems and, *a fortiori*, to non democratic structures. This is also due to national political influences such as the limitations imposed on decentralisation by various stakeholders, and to economic networks which are almost always national or international. As a result, The "localo-liberal" discourse (Rodary and Castellanet 2003) according to which the local would be the ideal form of natural resource commodification and the most suitable political organisation for biodiversity management, is more a matter of rhetoric than a reflection of reality on the ground (Pinton and Roué 2007, Rodary 2009).

These criticisms echoed another set of issues coming from ecologists, and particularly from powerful international NGOs, which favoured ecological and biological sciences, and relinquished poverty alleviation and livelihoods policies. The failure of sustainable management gave them new grounds to legitimate their political discourse (Brechin et al. 2003; Hutton et al. 2005). Indeed, according to these actors, a return to classic forms of conservation is justified first by the fact that local attempts at sustainable development likely to address biodiversity protection and livelihoods, have turned out to be unachievable; and second by the fact that the rate of destruction of the biological diversity requires urgent mobilisation. This political discourse amounts once again to excluding social issues from the sphere of conservation which is reasserted as a biological issue above all (Terborgh et al. 2002). The search for sustainable strategies at the local level was then abandoned in favour of more direct modes of investment associating ecological with economic efficiency, without direct social concern.

This reversal was clearly conveyed from the end of the 20th century, with the reassertion of the importance of outstanding natural areas, and with the rise of conservation NGOs in protection programmes. The three largest NGOs, i.e. the WWF, Conservation International (CI) and The Nature Conservancy (TNC), are increasingly marginalising local populations from the lands where they are conducting conservation programmes. The 'back to the barriers' approach and the re-establishment of the 'fortress conservation' for large scale conservation programmes, inevitably enhance the image of the large and powerful NGOs, since they are the only ones able to implement them, working at transnational or global level through hotspots, ecoregions and the like (see Rodary and Milian 2010). These regional or global policies reinforce the commodification of nature, for which globalised economic issues are becoming increasingly meaningful within conservation.

Outline of the Publication

Thus, for the past 30 years or so, the world of nature conservation has been experiencing important transformations in its objectives and operating modes. The main organisations dedicated to conservation have seen their means reinforced. Protected areas, as the main tools of conservation policies, continue to spread worldwide.

Today, protected areas around the world refer to a wide array of objectives, management models and legal statuses. They can offer complex forms of land use and resource planning or management: 'national parks', 'regional nature parks', 'protection areas', 'game reserves', 'biosphere reserves', 'agro-environmental measures', 'conservation networks', etc. Protected areas can also indicate the return of authoritarian policies legitimated by science. At the international level, the conceptualisation of protected areas has undergone three major evolutions that are expressed, in the best case scenario, concomitantly: the fact that human activities are taken into account, the constitution of transnational networks, and the extension of conservation issues to other sectors of activities. Their legal status linked to the pursuit of diversified objectives, their international networking and their transformation within large natural infrastructures confuses the very definition of protected areas. How is this expressed on the ground?

The choice of texts presented here aims to account for two simultaneous processes: the extension of the surface area of the protected areas, and the multiplication of their management methods. The aim is to determine whether the dynamics currently at work in the world of conservation, extend and reinforce former policies, or on the contrary, bring change, either via rupture, or through innovations likely to profoundly transform the modes of utilisation of nature. The book analyses the coherence between the definitions and the tools mobilised on the one hand, and the commitment of the actors on the other. The texts of this publication are developed in three main sections: How are protected areas redefined? Have new tools been mobilised? Have new territories been created?

These various contributions account for a certain continuity in the actions of conservation. In this regard, the emergence of sustainable development does not seem to have fundamentally modified operating modes, whether in the relations between local knowledge and scientific expertise, between scientific statements and political formulations, or yet in the organising modes between the local and the global. Although the current tendencies for returning to strict preservationist methods associated with the reinforcement of regional and international conservation policies and with international financing, are certainly modifying the tools conservationists have at their disposal, they are nonetheless prolonging a type of nature protection that was dominant during the whole of the 20th century.

Ruptures will more likely be found within the emergence of participation. For the past 20 years, the notion of participation has become the central theme of public intervention, present particularly in the process of widening and diversifying forms of conservation. Some associate it with the advance of democracy, others prefer to emphasise the community-based commitment to manage natural resources, while others only see ecological imperialism or a mere public relations exercise. The majority of the cases studied shows that the management of protected areas involves transformations and social connections as well as a conservationist injunction that both undermine any strong local participation. This does not mean that 'local' actors are deprived of their capacity for intervention. On the contrary, some of them find that conservation areas are places of training, spaces of negotiation or simply represent new opportunities. At this stage, the scientific challenge is to transcend participation to assess the way in which the collectivisation of conservation can foster innovative practices within local societies. Certain examples show that local re-appropriation is possible, even if that does not mean that local conflicts will be resolved or that the influence of market factors and transnational political systems will be reduced. Should these examples be verified, then sustainable development, through participation, will have caused a real innovation in the already long history of protected areas.

References

Adams W.M., 2004 – *Against extinction. The story of conservation.* London, Earthscan, p. 311.

Barreto Filho H. T., 2004 – "Notas para uma historia social das areas de proteção integral no Brasil". *In* Ricardo F. (ed.), *Terras indigenas e Unidades de conservação da natureza. O desafio das sobreposições.* São Paulo, Instituto socio-ambiental: 42–49.

Berdoulay V., Soubeyran O. (eds.), 2000 – *Milieu, colonisation et développement durable. Perspectives géographiques sur l'aménagement.* Paris-Montréal, L'Harmattan, p. 262.

Brechin S. R., Wilshusen P. R., Fortwangler C. L., West P. C. (eds.), 2003 – *Contested nature. Promoting international biodiversity with social justice in the twenty-first century*. New York, State Univ. of New York Press, p. 321.

Brosius J.P., Tsing A.L. et, Zerner C. (eds.), 2005 – *Communities and conservation. Histories and politics of community-based natural resource management*. Walnut Creek, AltaMira Press, p. 489.

Dumoulin D., Rodary E., 2005 – "Les ONG, au centre du secteur mondial de la conservation de la biodiversité". *In* Aubertin C. (ed.), *Représenter la nature ? ONG et biodiversité*. Paris, IRD Éditions: 59–98.

Ford C., 2004 – Nature, culture and conservation in France and her colonies 1840–1940. *Past & Present*, 183 (1): 173–198.

Grove R., 1995 – *Green imperialism. Colonial expansion, tropical island Edens and the origins of environmentalism 1600–1860*. Cambridge, Cambridge Univ. Press, p. 540.

Hutton J., Adams W. M., Murombedzi J. C., 2005 – Back to the barriers? Changing narratives in biodiversity. *Forum for Development Studies*, 2: 341–370.

IUCN, WWF, 1999 – *Principles and Guidelines on Indigenous and Traditional Peoples and Protected Areas*. Gland, UICN/WWF, p. 17.

IUCN, UNEP, WWF, 1980 – *World conservation strategy. Living resource conservation for sustainable development*. Gland/Nairobi, IUCN/UNEP/ WWF, p.72.

Larrère C., Larrère R., 1997 – *Du bon usage de la nature. Pour une philosophie de l'environnement*. Paris, Aubier, coll. Alto, p. 355.

McShane T. O., Wells M. P. (eds.), 2004 – *Getting biodiversity projects to work. Toward more effective conservation and development*. New York, Columbia Univ. Press, p. 442.

Milian J., Rodary E., 2010 – La conservation de la biodiversité par les outils de priorisation. Entre souci d'efficacité écologique et marchandisation. *Revue Tiers Monde*, 202: 33–56.

Nash R., 1967 – *Wilderness and the American mind*. New Haven, Yale Univ. Press, p. 256.

Pinton F., Roué M., 2007 – "Diversité biologique, diversité culturelle : enjeux autour des savoirs locaux". *In* Loyat J. (ed.), *Écosystèmes et sociétés. Concevoir une recherche pour un développement durable*. Paris, Quae/ IRD Éditions: 159–162.

Rodary E., Castellanet C., 2003 – "L'avenir de la conservation : du libéralisme local aux régulations transcalaires". *In* Rodary E., Castellanet C., Rossi G. (eds.), *Conservation de la nature et développement. L'intégration impossible ?* Paris, Karthala/GRET: 285–302.

Rodary E., 2009 – Mobilizing for nature in southern African community-based conservation policies, or the death of the local. Biodiversity and Conservation, 18 (10): 2585–2600.

Selmi A., 2006 – *Administrer la nature. Le parc national de la Vanoise*. Paris, Éditions de la MSH/Quae, p. 487.

Shackleton C. M. Willis T. J., Brown K., Polunin N. V. C. (eds.), 2010b– Community-based natural resource management (CBNRM): designing the next generation (Part 1). *Environmental Conservation, Special issue*, 37 (1): 1–106.

Shackleton C. M. Willis T. J., Brown K., Polunin N. V. C. (eds.), 2010b – Community-based natural resource management (CBNRM): designing the next generation (Part 2). *Environmental Conservation, Special issue*, 37 (3): 22–372.

Spiteri A., Nepal S. K., 2006 – Incentive-based conservation programs in developing countries: a review of some key issues and suggestions for improvements. *Environmental Management*, 37 (1): 1–14.

Terborgh J., van Schaik C., Davenport, L., Rao M. (eds.), 2002 – *Making parks work. Strategies for preserving tropical nature*. Washington, Island Press, p. 511.

Viard J., 1990 – *Le tiers espace. Essai sur la nature*. Paris, Méridiens Klincksieck, p. 152.

Vivien F.-D., 2005 – *Le développement soutenable*. Paris, La Découverte, coll. Repères, p. 122.

PART I
Redefining Protected Areas

In the first part of this publication, we address the definitions of protected areas in terms of spatial domains, land issues and categories of protection. These definitions pose a growing challenge in the context of the diversification of protected areas and the globalisation with which they are now inter-related.

By engaging with the World Database on Protected Areas of the World Conservation Monitoring Center, Estienne Rodary and Johan Milian conduct a detailed analysis of the spatial evolution and historical developments of protected areas. They examine whether the diversification of protection forms, particularly their integration into other forms of landscape utilisation, is a barometer of the possible ecologisation of development policies. Based on a cartographic analysis, their findings are mainly negative. The expansion of protected areas with sustainable management does not hide the significant fact that older forms of conservation are dominant, including highly protected areas. As a result, two processes are occurring simultaneously: the extension of sustainable forms of biodiversity management reflecting world politics, and a reinforcement of classic nature protection interventions focusing on the most outstanding sites and species. Thus, while it is possible for some sustainable management practices to be innovative (as shown in the third part of this publication), the world of conservation seems to rely partially on more traditional modes of protection, not only of nature, but also of its own institutional evolution.

In conservation biology, the classic insular model is progressively giving way to a more complex and diversified reticular model which can take into account connectivity, but which is more delicate in terms of management. This is manifest with Marine Protected Areas (MPAs) which have been increasing rapidly in numbers and surface areas for about twenty years, particularly in the inter-tropical zone. Even if MPAs originate in the biological disciplines, it appears that, as conservation and sustainable development tools, they cannot be dealt with by referring only to the biological and environmental dimensions. Using their scientific disciplines, Christian Chaboud, Florence Galletti, Ambroise Brenier, Gilbert David, Philippe Méral, Fano Andriamahefazafy and Jocelyne Ferraris examine the biological, legal, economic and geographic particularities of marine protected areas and show the resulting need to take into consideration coastal areas, sea surfaces, sea beds, territorial waters and open seas in managing MPAs. What is important is not so much the study of the MPA as a specific object under construction, but its particularities in terms of governance. The current MPA race and its corollary, the 'race for efficient governance', mean that scientific disciplines are increasingly being used in the study, implementation and evaluation of MPAs.

While it is difficult to speak of rupture in the forms of conservation, a clear evolution is emerging that presents issues of management, governance between public and private bodies, as well as lands as multidisciplinary and experimental domains, on large scales, and associating different types of protected areas under a common policy. In turn, these reconfigurations have institutional effects leading to the redefinition not only of protected areas, but also of various groups of actors and experts working on them.

Chapter 1

Expansion and Diversification of Protected Areas: Rupture or Continuity?

Estienne Rodary and Johan Milian

These past 30 years, protected areas have experienced a worldwide dramatic increase in their surface area and have been profoundly influenced by the discourses on globalisation and sustainable development. They have become an integral part of a wider framework of environmental policies reaching beyond the physical boundaries of conservation spaces. The dynamics of expansion and diversification potentially best reflect the orientation of current biodiversity conservation policies and protected areas evolution. This chapter attempts to test the validity of these two dynamics of expansion and diversification by using the World Database on Protected Areas (WDPA). Its main hypothesis is that changes in conservation policies can be illustrated by changes in the number, type and surface area of protected areas.

It is well known that *in situ* biodiversity conservation has drastically expanded over the past 30 years, and in fact for more than 100 years. With almost 20 million km² currently under conservation policies, protected areas represent one of the main modes of land tenure worldwide. In this context, protected areas management now embraces the debate around sustainable development and its effective implementation.

Because this rhetoric of expansion and diversification towards sustainable development is scientifically and politically important, this chapter confronts it with an analysis of the most comprehensive information available. The WDPA[1] is managed by the World Conservation Monitoring Center (WCMC) within the United Nations Environment Programme (UNEP), in partnership with the World Commission on Protected Areas of the International Union for Conservation of Nature (IUCN-WCPA). It benefits from the input of most of the major conservation NGOs. It is also used for the four-yearly publication of the United Nations' list of protected areas[2]. The WDPA is the only exhaustive global compilation of its kind concerning protected areas. It utilises classifications based on a categorisation system proposed by the IUCN, which aims to allow conservation regulations within different national systems to be compared internationally.

---

1   http://www.wdpa.org.
2   See Chape et al. (2003) and IUCN (1998), for the last two publications of the official list of UN protected areas.

This chapter examines current developments in conservation based on the results of processing the WDPA. It shows that, although conservation organisations claim that the rate of creation of protected areas is on the increase, it has in fact been declining in recent years, a possible indication that the golden age of protected areas is now behind us. This could be linked to the diversification of protected areas, which despite mainstream discourse, only marginally involves sustainable forms of nature management, even though these forms could reflect the most innovative management that will define tomorrow's conservation.

The Current Production of Protected Area

For almost 30 years, the world of conservation has been influenced by expansion and diversification. Each trend has sought to overcome various issues as perceived by conservationists, and to give new legitimacy to their practices, in particular in the context of the evolution of environmental international politics (cf. the introduction of this book). The community-based and local participation approach has probably been the most significant change during this period. The move backwards towards fortress conservation policies has partially operated against this community-based policy and its presumed limitations to advance its own agenda. It is within this general framework of divergent political options that conservation experts have increased the number of tools for defining and creating protected areas, while also broadening their scope. First, by identifying both strategically and globally, zones considered to be a high priority for conservation, and with the potential to be classified as protected areas. Second, by seeking a more comprehensive definition of a nature protected area, particularly through the extensive inclusion of less protected, 'sustainable' areas. And finally by allowing private and community-based spaces to be classified as protected areas (cf. the various examples given in Part III of this book).

The Strategic Spatial Approaches

The integrative approach which focuses on community participation and the defensive approach which centres on fortress conservation share the view that there is still a requirement to increase the surface of protected areas. The former includes a spatial dimension in which the co-management is generally carried out either on the periphery of existing protected areas, or within protected areas that had not benefited from effective management plans up to that point (e.g. the 'paper parks'). As such, the participation policies are as much a method of integration as they are a set of tactics for pushing back the interface between the wilderness and society. The latter, while promoting a movement 'back to the barriers', is also developing strategies for the expansion of conservation areas. In this sense, both approaches share a similar attempt to define future endeavours by looking beyond the status quo (Margules & Pressey 2000; Pressey et al. 2007). This attempt is mainly pursued at two levels. First at the regional level, where the 'ecosystem' approach, initiated by the 1992 Convention on Biological Diversity, which aims to integrate protected areas and surrounding lands (i.e. agricultural, etc.) into a join management plan, is now a prerequisite in all biodiversity management policies (Smith & Maltby 2003). Second, at the global level, where major conservation organisations create protected areas using priority regions defined according to various ecological criteria (endemism, diversity, endangered species, etc.). In this regard, the Conservation International's Biodiversity Hotspots, Birdlife International's Important Bird Areas, the Global 200 ecoregions initiated by the WWF and the portfolios of The Nature Conservancy, compete to shape the actions of conservation worldwide … as far as ecological priorities and media coverage are concerned[3].

The Compilation Tools for Conservation

Moreover, the globalised approach has been partly based on the compilation and harmonisation of the tools available to evaluate conservation, which have taken on a more inclusive dimension over the last decade. The IUCN has been at the centre of this undertaking since the 1960s, when the First World Congress on protected areas took place and the first protected area classifications were established. Although IUCN categorisation is not binding on national legislations, it has an obvious impact on the clarity and legitimacy of conservation policy. Since the 4th World Congress on National Parks and Protected Areas held in Caracas in 1992, the IUCN has defined seven categories of protection measures (I to VI), with the first category made up of two sub-categories (Ia and Ib) (See Table 1.1).

More recently, the Commission on Environmental, Economic and Social Policies of the IUCN launched a new phase in the expansion of protected areas, by proposing to take into account a typology outlining modes of governance. This means that protected areas are no longer distinguished simply by their degree of protection, but also in terms of the institutions managing them (states as well as private and community-based actors) (See Table 1.1). It is within this framework that new types of protected areas are now included in the WDPA, although these categories are not always filled in properly. Today, the options for designating protected areas favour an inclusive approach indicative of the diversification policies currently being followed, although with clear controversy between experts, as witnessed with the recent cautious stand adopted by IUCN (see Dudley 2008). Finally, this classification work is associated with programmes conducted within the WCMC to expand and modernise the WDPA.

---

3   See Redford et al. (2003) for a list of these methods and Rodary and Milian (2010) for a critical review.

Table 1.1   The protected area matrix

Protected Area Categories

Ia. Strict Nature Reserve	Ib. Wilderness Area	II. National Park	III. Natural Monument or Feature	IV. Habitat/Species Management Area	V. Protected Landscape/ Seascape	VI. Protected area with sustainable use of natural resources	No Category
Integral protection, for scientific research	Integral protection	High protection, but tourism authorised	Protection restricted to a specific site	Protection concerns a specific species or ecosystem	Low protection, transformed environment	Low protection, sustainable use of resources	Un-determined

Governance Types

A. Government			B. Shared			C. Private			D. Indigenous and local communities	
State department or agency	De-centralised government	Other institution (de-legation)	Trans-boundary management	Collaborative management (unique structure with external influence)	Joint-management (different partners)	Individual owner	Non-profit organisation	Commercial organisation	Indigenous peoples	Local communities (sedentary and mobile)

Source: Dudley 2008

Expansion or Diversification: A Cartographic Analysis

The arguments over the opening or closing of conservation to other forms of land use, and the attempts to integrate information and management tools on a global scale, are orientating the categorisation of protected areas. The cartographic analysis presented here makes it possible to evaluate these orientations in regards to the initial hypotheses of protected area expansion and diversification. We present at first the methodological framework that structured our approach, followed by an analysis of the trends of expansion and diversification.

Methodological Considerations

Free access to the WDPA via the Internet has been possible since 2003. At the time, it represented an important novelty for all those who specialised in protected area management. It meant being able to obtain information on a specific protected area or a state and, more importantly, to carry out comparisons at a national or international level. Because attitudes towards conservation are also political choices and not just the application of a supposedly neutral science, analytical and management tools have been at the centre of important arguments. As a result, there is, on one hand, a manifest shortage of studies on the current state of conservation, and of comparative studies on a global scale in specific domains (particularly social sciences)[4], and, on the other hand, a proliferation of studies that exceed mere scientific analyses.

We can easily imagine that, for political reasons mainly[5], the partners of the WDPA project might use the database without necessarily wishing to publish in detail the results of their studies or use their publications for their own promotion. Nevertheless, it is hard to understand why researchers working on conservation issues appear to ignore altogether this type of tool, if not for being too focused on local examples, thereby neglecting global analyses (Rodary 2009).

A large number of WDPA-based studies have been published, but they have been mainly interested in gap analyses between protected areas and ecological conditions, apart from the studies produced directly by WCMC or IUCN experts

---

4   With an increasing number of exceptions nonetheless. For examples that are not mere compilation of case studies, see James et al. (1999); Redford et al. (2003); Agrawal and Redford (2006); Halpern et al. (2006); Hayes (2006); Naidoo et al. (2006); Depraz (2008); Héritier and Laslaz (2008); Leverington et al. (2008), and more significantly, the recent endeavour by D. Brockington and his colleagues to engage in a global understanding of the social incidences of conservation policies (Brockington et al., 2008).

5   There is a particularly appropriate example of this in the debate created by M. Chapin (2004). In response to his article showing the decline in the collaboration between conservationist and indigenous movements, major conservation organisations – all members of the WDPA consortium – simply gave local examples, making sure not to supply figures of the impacts their policies had at the global level (Collective 2005).

which mainly show the current conditions under which protected areas are distributed, and which are rather biased in favour of the promotion of the WDPA (Green and Paine 1998; Chape et al. 2005; Chape et al. 2008; Hoekstra et al., 2010)[6]. Adopting a more biological stance, the various conservation organisations promoting the prioritisation methodologies, carried out analyses using the WDPA to evaluate the importance of specific biomes or ecosystems. While these analyses give neither specific historical, nor social information and focus on ecological criteria (see Rodary and Milian 2010), they nevertheless comprise the first global analyses that measure the pertinence of the current distribution of protected areas for species and natural environments[7].

Because processing a database leads to an inevitable loss of information, the analyses carried out here are not a description of the reality on the ground but, more simply, an analytical compilation of national figures. Our conclusions must be placed in this specific methodological context. The analyse focuses on two particular themes: the historic development of protected areas and the use of protected area management categories. On both points however, the WDPA reveals limitations: on the one hand the database either contains errors or has not been updated on time where information willingly supplied to the WCMC by national authorities or conservation managers is concerned. We decided for the sake of coherence to keep the database 'as is', even if we were in possession of more recent information for specific regions. On the other hand, not all the attributes of each database entity (i.e. protected area) are filled in. In particular, surface data (and a *fortiori* georeferencing) as well as the date of creation, are not systematically indicated (See Box 1.1).

Trends in Expansion

Figure 1.1 illustrates the evolution of the total surface area of protected areas since 1870, a symbolic date when the first American national park was created (for a general history, see Adams 2004). What can be seen from this graph is obvious: throughout the 20th century, there has been major growth in both the size and importance of protected areas. When considering the way man has defined much of the earth's surface, the 20th century has not only been the century of urbanisation, but also that of the institutionalisation of protected areas as an instrument of environmental management. From 68,000 km² in 1900 to 1 million km² in 1950, the

---

6   But note the article of K. Zimmerer and his colleagues which proposes an analysis of the historical development of protected areas between 1985 and 1997 (Zimmerer et al. 2004), and also an older article by IUCN experts (Harrison et al. 1982).

7   On this theme of full development, of note are the articles of Myers et al. (2000), Olson and Dinerstein (2002), Rodrigues et al. (2004), Brooks et al. (2004), Hoekstra et al. (2005); Kareiva and Marvier (2007); Langhammer et al. (2007) and Pyke (2007) among many others. For an analysis of these issues on prioritisation, see our article Rodary and Milian 2010.

Box 1.1   Methodological limitations

As a tool, the WDPA has a certain number of technical and informational limitations which we have had to take into account throughout our analysis.

The data for surface area is not available for a limited number of protected areas: 12.3% of the sites listed in the database have been given a value of zero for the surface area field. In the majority of cases, these are protected areas with a surface area of less than 1 ha, most of which are classified under Category III. However, in a small number of cases, these are probably protected areas with larger surface areas. But since this shortcoming is rare, it does not result in a significant margin of error for the results derived from this variable.

We were not able to identify the date of creation for 22% of the protected areas. Concerning Category Ia to VI sites, calculations have been carried out on 65% of the sites. This absence of data concerns a small number of countries, mainly Russia and New Zealand (which represent 50% and 17% respectively of sites with this problem). If we exclude these two countries, only 11.55% of the sites lack values concerning their date of creation. For those protected areas having a degree of protection that has not been evaluated by the IUCN ('no category', cf. hereunder), calculations have been carried out on 51.7% of the sites. Then again, considering their typology, many of these sites probably have a surface area of less than 1 ha.

'000 000 km^2

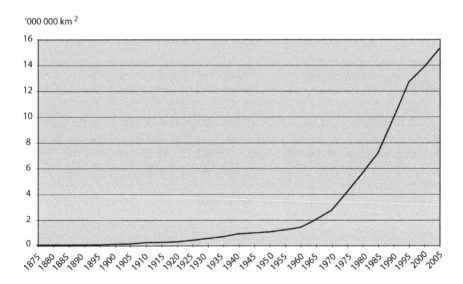

Figure 1.1   Evolution of the total surface area of protected areas worldwide

surface area has multiplied more than tenfold in 50 years, and has grown by a factor of fifteen over the second half of the century, to reach more than 15 million km^2 in 2005 (19.33 million km^2 if we take into account areas with an unknown date of creation). Nonetheless, we can also observe a slight inflexion in the rate of increase, dating from 1995. The highest rate of increase in fact took place during the previous decade, i.e. between 1985 and 1995.

Figure 1.2, illustrates surface areas placed under protection every five years. We can clearly distinguish a slowing down in the creation rate of protected areas since 1995, which was still an important period of growth but was lower than the progress observed between 1975 and 1995.

'000 000 km^2

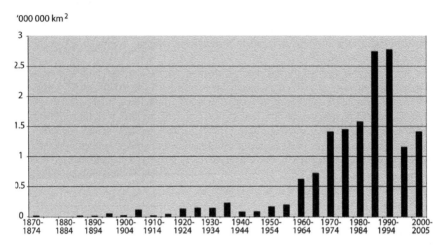

Figure 1.2   Five-year evolution of surface areas under protection worldwide

Plate 1 examines the rate of expansion of protected areas by continent[8]. An initial analysis shows the marked dominance of the North American and African continents. It is not until the 1960s that other continents experienced a noticeable establishment of protected areas. From the 1960s onwards, this is very conspicuous for Asia and South America; from 1975 for Oceania; and from 1985 for the Middle East. After these dates, there is a striking degree of similarity in creation rates across the continents, with the exception of Europe where it has remained constant, though moderate, since 1960.

Trends in Diversification

A reading of the figures by category rather than continent deepens the analysis of recent dynamics in conservation. Plate 2 shows the development of the number of protected areas worldwide since 1900 arranged by categories. Unsurprisingly, Categories III and IV are predominant over the entire period. Category III concerns small protected areas, very often less than 10 km², whereas Category

---

8   Here we created groups that do not match those used by the WCMC to make the comparisons clearer.

IV includes those areas dedicated to the management of a specific species, or group of species, i.e. animal reserves, game reserves or forest reserves. Up until the 1930s, these reserves represented the main type of protection, and to this day remain the preferred measure when protection is to be ensured through the targeted use of specific natural resources. It comes as no surprise that Category II is represented by only a few sites; national parks usually cover vast territories, which mechanically restricts their numbers (Milian and Rodary 2008). Perhaps more unexpected however, is the low proportion of sites in Category VI: supposedly to represent the paradigmatic form of new policies integrating both conservation and development, these areas are still marginal compared to the other categories. Categories V and VI which were conceived explicitly with sustainable development in mind, constitute little more than 10% of world sites.

Also noteworthy is the parallel increase in the number of sites across almost all categories; with the exception of Category IV, for more than 40 years all those categories historically registered by the IUCN have been following comparable development rates, this being true even for Category Ia which is highly protectionist and forbids any utilisation of natural resources.

'No category' sites also represent an important part of the total number of protected areas: these are spaces with an unknown management strategy, including territories that do not appear in the United Nations list because of their modes of governance. Referring to the designations given by governments to these areas can give us an idea of the major classifications concerned. What emerges from Table 1.2 is that indigenous reserves and forest reserves constitute more than half of the surface area concerned. The former are the direct consequence of community-based policies developed in the 1990s. The latter illustrates the IUCN and the WCMC's desire to integrate types of managed areas that national legislations do not necessarily categorise as tools for conservation. South America (54%), Africa (21%) and Asia (16%) are the continents the most affected by these 'no category' sites. In fact, 99% (i.e. more than 1 million km²) of indigenous reserves are found in South America and close to 60% (i.e. 348,000 km²) of forest reserves in Africa.

Plate 3 also shows different categories of protected areas although this time they are classified according to their respective surface areas on a global scale. Here the orders are reversed: Categories II and VI, on their own, represent more than half of the total surface area of protected areas. The increase has been particularly marked for "areas with sustainable management of resources" (VI)[9].

---

9   In the WDPA, although Category VI has been registered since the beginning of the 20th century, it actually represents a retrospective categorisation as it was only created at the beginning of the 1990s, and is therefore anachronistic. It highlights nonetheless the existence of 'sustainable' forms of management throughout the century, although it is only since the 1970s that this method of management became obvious geographically. This anachronism enables contemporary conservationists to classify under Category VI reserves

Table 1.2   Classification of 'no category' protected areas

Indigenous Reserves	36.60%
Forest Reserves	19.64
Nature Reserves	8.91
National Parks	8.10
Nature Parks	4.13
Agriculture	3.40
Animals Reserves	2.95
Marine Areas	2.85
Game Reserves	2.62
Monuments	2.24
Wetlands	0.78
Habitats	0.13
Recreation Areas	0.07
Unknown Designation	7.58
Total	100.00

It was also the case for national parks (II) which are generally far removed from sustainable management. It follows that the current dynamic behind the expansion of protected areas cannot be explained entirely by the diffusion of the least protected spaces or, to put it another way, the most integrated spaces. However, we must note that during the last decade when the growth rate decreased compared to the previous period, only 'sustainable categories' experienced a reinforcement of their numbers.

Box 1.2 contains a summary of the main characteristics of protected areas, and compares their surface areas by continent. It also looks at their spatial and geographical evolution as well as status.

A Historical Geographical Synthesis, or How to Conserve Conservation

The Plates 4 to 10 summarise the historical evolution rates of protected areas, the level of their protection and their locations. For all the maps, three historical periods have been defined, each corresponding to major development phases in conservation; and the categories of protection have been represented in two groups[10]. We propose an

---

that would be more appropriately classified under Category IV (Habitat and Species Management Area, corresponding to hunting or forest reserves).

10   The three historical periods are pre-1970, 1970 to 1985 and post-1985. The extent of protection refers to Highly Protected (Categories Ia to IV) and Sustainable (Categories V

Box 1.2   Main data on protected areas

Protected areas by continent

Continent	Number	Surface Area (in km²)	% of the continent[1]
North America[2]	11,669	3,876,180	17.79
Africa	5,897	3,041,052	10.04
South America[3]	3,904	3,827,243	18.82
Asia	8,273	4,155,537	11.31
Oceania	10,171	1,894,610	21.18
Europe	57,493	1,296,395	12.47
Middle East[4]	786	1,158,365	16.38
Total	**98,193**	**19,249,382**	

(1) Percentage given for information only, considering that a small proportion of protected areas is composed of marine areas.
(2) Including Greenland but excluding Mexico.
(3) Including Mexico, Central America and the Caribbean.
(4) Including Afghanistan and excluding Egypt attached to Africa.

Surface of protected areas

- Constant progression since 1870.
- Historical precedence of the North American and African continents.
- Slowing down in the rate of creation since 1995.
- Predominance of Categories II (National Park) and VI (Managed Resource Protected Areas).

Number of protected areas

- Predominance of Categories III (Natural Monument) and IV (Habitat and Species Management Area).
- In non-categorised protected areas, predominance of indigenous reserves.

interpretation of the maps by continent and according to three modes: by historical phase, zones of high concentration, and categories of protection.

Historical Progression

In terms of historical progression, we can distinguish seven major development phases in conservation over 125 years:

and VI). Also, the use of dots (as mentioned in the legend) does not refer to small protected areas but to units that are not georeferenced (Box 1).

- An *institutionalisation phase* between 1870 and 1920, which corresponds to the first measures of protection. At this time, most protected areas are either reserve or national park, and are predominantly found in Africa and North America respectively – the two regions most affected by early conservation policies. But interest in conservation begins to spread to other regions of the world, certain British dominions in particular such as New Zealand, to Scandinavian and Western Europe (the first national parks appeared in Sweden, Switzerland and Spain; nature reserves appeared also in the Netherlands, Denmark and Sweden) and to South America (first Chilean reserves).

- A *progression phase* between 1920 and 1940, in line with the consolidation of conservation policies in Africa and North America, and with the associated dissemination of the national park as the main tool for protection worldwide (Japan, Europe and the southern Cone of South America). As the creation of nature reserves spreads throughout Europe, other regions too become aware of conservation, particularly with the creation of wildlife reserves in Asian British dominions (i.e. India, Burma and Sri Lanka), as well as forest reserves in the Dutch East Indies.

- A *slowing down phase* between 1940 and 1960, as a result of the Second World War and decolonisation, with however a few noticeable expansions, in Australia and New Zealand in particular. Yet it is during this period that the structures which will really boost the internationalisation of conservation are going to be set up, such as the IUCN and the WWF. For this reason, the recession phase for the creation of protected areas corresponds to a reconfiguration period, where political frameworks change with decolonisation, and where the economic context also changes with the internationalisation and democratisation of nature tourism.

- A *resumption phase* between 1960 and 1970, during which conservation policies are taken over by post-independence governments. Partially because of this political change, the legitimacy of protected areas shifts towards an economic discourse mainly connected to the development of the tourism industry. During that period, South America and, to a lesser extent Europe and Africa, experience a very clear acceleration in the creation of protected areas.

- A *strong progression phase* between 1970 and 1985, corresponding to the appearance of environmental issues on the international political agenda – what has been called the emergence of global 'eco-politics' (Le Prestre 2005). This phase corresponds to the real globalisation of protected areas as the dominant tool for conservation policies. In this regard, there was strong progress on continents that, until then, were little affected by this movement (Asia, insular Oceania and to a lesser extent the Middle East). Low-density areas also benefited from large-scale protection (Alaska, Northwest Canada, Greenland, Arctic Siberia and the Kunlun Massif).

- An *intensification phase* between 1985 and 1995, during which all the continents, without exception, experience their highest rate of area protection, associated with the institutionalisation of sustainable development as an international discourse. This phase sees the creation of large Category II parks (the Tassili N'Ajjer National Park and the Ténéré National Nature Reserve in the Sahara; in the Amazon and in the Chaco in South America), as well as the creation of big protected areas of Category V (Tibetan Plateau and the Himalayas) and Category VI (Arabia, Central Australia, Quebec and Argentina). It is during this period that the great majority of governments institute public policies for biodiversity protection, and commit themselves internationally to conservation.

- Finally, a *slowing down phase* between 1995 and 2005, during which several continents experience a decrease in their rate of protected area creation (even though the pace remains sustained in Asia – in China in particular – Oceania and Europe). This occurs as conservation policies come under increasing scrutiny, both at local level with regard to community participation, and globally with the decline of states' commitment towards multilateral agreements on environmental issues (Rodary 2007).

Spatial Configuration

In terms of distribution, the current situation shows a globalisation of the instruments for spatial protection, yet with some marked differences across country and continent. On a regional scale, we can distinguish three main types of configuration in the larger concentrations of protected areas:

- *Very large blocs of protected areas*. These are situated predominantly in the polar or circumpolar areas (Antarctica[11], Greenland, Northern Canada, Alaska, Southern Chile and Siberia) and the great deserts (Arabia, Sahara, Namib, Kalahari, Tibetan High Plateau and Xinjiang as well as Mongolia).
- *Networks of protected areas of lesser importance*. These are found in the circum-Amazonian and Andean area, in Central America, on the Australian coasts, in Eastern Africa and in the Indo-Malaysian Archipelago.
- *Regions with high concentrations of small management units*. These are located mainly in highly urbanised and/or densely populated areas: North-East United States, East-Central Europe, Brazilian coasts, Eastern China, Japan, Korea and India.

In terms of protection categories, there are marked differences by continent (See Table 1.3). On the one hand, all the continents have a special category that represents more than one third of the surface area under protection. On the other hand, we see an obvious relationship between these favoured categories and the

---

11   Although the Antarctic continent does not appear in the list, it is fully protected by the Protocol on Environmental Protection to the Antarctic Treaty since 1997.

Table 1.3   Main protection category by continent

Continent	Main category in surface area	Percentage of protected surface area
Middle East	VI. Managed Resource Protected Area	76.21
Oceania	VI. Managed Resource Protected Area	50.09
North America	II. National Park	42.56
South America	'No Category' (Indigenous Reserve)	42.43
Europe	IV. Habitat/Species Management Area	35.87
Asia	V. Protected Landscape	35.60
Africa	II. National Park	33.78

main development period of protected surface areas on the continent considered (Plate 1).

From this synthesis one could infer that continents which created conservation areas much later on, favoured sustainable forms of management. But although this analysis is correct for the Middle East and Oceania (even if the data is partially distorted by the creation of some very large units, but which is little representative of the most widespread categories in the region), it indicates above all the strong permanence of older protected areas.

Conclusion

Since the emergence of sustainable development as a dominant discourse, the world of conservation managed to 'conserve' its own means of action, organised mainly around protected areas. Admittedly, some of these areas are becoming increasingly connected to other spaces, integrated with other types of territorial management and have been diversified by giving a more important role to human practices not directed explicitly towards biodiversity conservation. Yet at the same time, there has been an expansion of the more traditional protected areas, which invalidates the perception that 'conservation' has lapsed into 'sustainable development' by having given up the specificities that have defined and founded protected areas for more than 100 years.

The expansion of protected areas has been confirmed in the last few decades, during which even the most marginal states (as far as conservation policies are concerned), have embarked on the creation of protected areas. What we have been dealing with for the past 30 years is the globalisation of this tool, even if the regional differences in surface area and category of protection remain very marked. Does a slowing down of the creation rate of protected areas, as observed in recent years, reflect a rupture in the dynamics of expansion worldwide? Although it is still too early to provide a clear-cut answer, we can envisage that the future of conservation will be characterised by a double-faced trend. One is the consolidation of existing land tenure systems of protection – and therefore the perpetuation of current conservation policies. The other is the development of experimental and in future more innovative approaches, mainly by marginal organisations in the field of conservation policies. This double trend enables conservation to respond to the constraints of the time, while protecting the conservation legacy.

References

Agrawal A., Redford K., 2006 – *Poverty, development, and biodiversity conservation: shooting in the dark?* New York, Wildlife Conservation Society, Working paper n° 26, p.50.

Brockington D., Duffy R., Igoe J., 2008 – *Nature unbound. Conservation, capitalism and the future of protected areas.* London, Earthscan, p. 240

Brooks T. M., Bakarr M. I., Boucher T., da Fonseca G. A. B., Hilton-Taylor C., Hoekstra J. M., Moritz T., Olivieri S., Parrish J., Pressey R. L., Rodrigues A. S. L., Sechrest W., Stattersfield A., Strahm W., Stuart S. N., 2004 – Coverage provided by the global protected-area system: is it enough? *Bioscience,* 54 (12): 1081–1091.

Chape S., Blyth S., Fish L., Fox P., Spalding M., 2003 – *2003 United Nations List of Protected Areas.* Gland/Cambridge, p. 44.

Chape S., Harrison J., Spalding M., Lysenko I., 2005 – Measuring the extent and effectiveness of protected areas as an indicator for meeting global biodiversity targets. *Philosophical Transactions of the Royal Society B* (360): 443–455.

Chape S., Spalding M.D., Jenkins M.D. (eds.), 2008 – *The world's protected areas. Status, values and prospects in the 21st century.* Berkeley, University of California Press, p. 376.

Chapin M., 2004 – A challenge to conservationists. *Worldwatch Magazine,* November–December: 17–31.

Collective, 2005 – A challenge to conservationists: Phase II. From reader. *Worldwatch Magazine,* January–February: 5–20.

Depraz S., 2008 – *Géographie des espaces naturels protégés. Genèse, principes et enjeux territoriaux.* Paris, Armand Colin, coll. U Géographie, p. 320.

Green M. J. B., Paine J., 1998 – "State of the world's protected areas at the end of the twentieth century." *In IUCN, Protected areas in the 21st century. From islands to networks,* Albany, 23–29 November 1997, IUCN: 104–126.

Halpern B. S., Pyke C. R., Fox H. E., Haney J. C., Schlaepfer M. A., Zaradic P., 2006 – Gaps and mismatches between global conservation priorities and spending. *Conservation Biology,* 20 (1): 56–64.

Harrison J., Miller K., McNeely J. A., 1982 – The world coverage of protected areas: development goals and environmental needs. *Ambio,* 11 (4): 238–245.

Hayes T. M., 2006 – Parks, people, and forest protection: an institutional assessment of the effectiveness of protected areas. *World Development,* 34 (12): 2064–2075.

Héritier S., Laslaz L. (eds.), 2008 – *Les parcs nationaux dans le monde. Protection, gestion et développement durable*. Paris, Ellipses, coll. Carrefours Les dossiers, p. 312.

Hoekstra J. M., Boucher T. M., Ricketts T. H., Roberts C., 2005 – Confronting a biome crisis: global disparities of habitat loss and protection. *Ecology Letters*, 8 (1): 23–29.

Hoekstra J., Molnar J. L., Jennings M., Revenga C., Spalding M. D., Boucher T. M., Robertson J. C., Heibel T. J., Ellison K., 2010 – *The atlas of global conservation. Changes, challenges, and opportunities to make a difference*. Berkeley, University of California Press, p. 272.

James A. N., Green M. J. B., Paine J. R., 1999 – *A global review of protected area budgets and staff*. Cambridge, WCMC/World Conservation Press, WCMC Biodiversity Series n° 10, p. 55.

Kareiva P., Marvier M., 2007 – *Conservation for the people, Scientific American*, 297 (4): 50–57.

Langhammer P. F., Bakarr M. I., Bennun L. A., Brooks T. M., Clay R. P., Darwall W., De Silva N., Edgar G. J., Eken G., Fishpool L. D. C., da Fonseca G. A. B., Foster M. N., Knox D. H., Matiku P., Radford E. A., Rodrigues A. S. L., Salaman P., Sechrest W., Tordoff A. W., 2007 – *Identification and gap analysis of key biodiversity areas. Targets for comprehensive protected area systems*. Gland, IUCN, p. 116.

Le Prestre P., 2005 – *Protection de l'environnement et relations internationales: les défis de l'écopolitique mondiale*. Paris, A. Colin, p. 477.

Leverington F., Hockings M., Costa K.L. (eds.), 2008 – *Management effectiveness evaluation in protected areas – a global study*. Gatton/Gland/Washington, The University of Queensland/IUCN WCPA/TNC/WWF, p. 70.

Margules C. R., Pressey R. L., 2000 – Systematic conservation planning. *Nature*, 405 (6783): 243–253.

Milian J., Rodary E., 2008 – "Les parcs nationaux dans le monde, un aperçu cartographique". *In* Héritier S., Laslaz L. (eds.), *Les parcs nationaux dans le monde. Protection, gestion et développement durable*. Paris, Ellipses, coll. Carrefours Les dossiers: 33–44.

Myers N., Mittermeier R. A., Mittermeier C. G., da Fonseca G. A. B., Kent J., 2000 – Biodiversity hotspots for conservation priorities. *Nature*, 403 (6772): 853–858.

Naidoo R. et al., 2006 – Integrating economic costs into conservation planning. *Trends in Ecology & Evolution*, 21 (12): 681–687.

Olson D. M., Dinerstein E., 2002 – The Global 200: priority ecoregions for global conservation. *Annals of the Missouri Botanical Garden*, 89 (2): 199–224.

Pressey R.L. et al., 2007 – *Conservation planning in a changing world. Trends in Ecology & Evolution*, 22 (11): 583–592.

Pyke C. R., 2007 – The implications of global priorities for biodiversity and ecosystem services associated with protected areas. *Ecology and Society*, 12 (1): http://www.ecologyandsociety.org/vol12/iss11/art14/.

Redford K. H., Coppolillo P., Sanderson E. W., da Fonseca G. A. B., Dinerstein E., Groves C., Mace G., Maginnis S., Mittermeier R. A., Noss R., Olson D., Robinson J. G., Vedder A., Wright M., 2003 – Mapping the conservation landscape. *Conservation Biology*, 17 (1): 116–131.

Rodary E., 2007 – "La gouvernance de la biodiversité et le développement". *In* Jacquet P., Tubiana L. (eds.), *Regards sur la Terre. Biodiversité, nature et développement*. Paris, Presses de Science Po/AFD: 137–152.

Rodary E., 2009 – Mobilizing for nature in southern African community-based conservation policies, or the death of the local. *Biodiversity and Conservation*, 18 (10): 2585–2600.

Milian J., Rodary E., 2010 – La conservation de la biodiversité par les outils de priorisation. Entre souci d'efficacité écologique et marchandisation. *Revue Tiers Monde*, (202): 33–56.

Rodrigues A. S. L., Andelman S. J., Bakarr M. I., Boitani L., Brooks T. M., Cowling R. M., Fishpool L. D. C., da Fonseca, G. A. B., Gaston K. J., Hoffmann M., Long J. S., Marquet P. A., Pilgrim J. D., Pressey R. L., Schipper J., Sechrest W., Stuart S. N., Underhill L. G., Waller R. W., Watts M. E. J., Yan X., 2004 – Effectiveness of the global protected area network in representing species diversity. *Nature*, 428 (6983): 640–643.

Smith R. D., Maltby E., 2003 – *Using the ecosystem approach to implement the Convention on Biological Diversity. Key issues and case studies*. Gland/Cambridge, IUCN, p. 118.

UICN, 1998 – *1997 United Nations list of protected areas*. Gland/Cambridge, UICN, p. 412.

Zimmerer K., Galt R. E., Buck M. V., 2004 – Globalization and multi-spatial trends in the coverage of protected-area conservation (1980–2000). *Ambio*, 33 (8): 520–529.

Chapter 2

Marine Protected Areas and Governance: Towards a Multidisciplinary Approach

Christian Chaboud, Florence Galletti, Gilbert David, Ambroise Brenier,
Philippe Méral, Fano Andriamahefazafy and Jocelyne Ferraris

Marine Protected Areas (MPAs) occupy a special place among the actors and processes involved in the governance of protected areas. Although according to the current definition of the IUCN, MPAs apply only to marine areas[1], in this chapter we will also include in this category Marine and Coastal Protected Areas (MCPAs), which encompass both marine and terrestrial components.

Over the past 30 years, the number of marine protected areas has been increasing rapidly in inter-tropical areas, where biodiversity conservation challenges are greater than anywhere else. This number rose from 118 in 1970 to 319 in 1980 (Silva et al., 1986), and eventually exceeded 1300 in 1995 (Kelleher et al. 1995). Of these, 400 MPAs concern coral reefs exclusively (Salvat et al. 2002). In 2003, during the World Parks Congress organised in Durban by the IUCN, a resolution was adopted to classify 20% of the world's seas as MPAs within 20 to 30 years' time.

Island states are particularly involved in implementing this resolution. In Oceania, in 2005, the Fiji government undertook to convert 30% of its exclusive economic area into MPAs by 2020. In 2006, the governments of Palau, Guam, the Federated States of Micronesia, the Northern Mariana Islands and the Marshall Islands pledged to follow similar objectives within the framework of the Micronesian Challenge. In the Indian Ocean, the Indian Ocean Commission is in the process of implementing a regional network of MPAs while the government of Madagascar has also embarked on an ambitious programme to create MPAs.

While MPAs are still limited in absolute surface area compared to their terrestrial counterparts (Rodary and Milian, this publication), their expansion continues at a rapid pace. In 2005, there were 5127 marine protected areas (including 967 at an international level) representing 0.6% of the surface area of the oceans. The French government is promoting the creation of new MPAs, not only in France, but particularly in its overseas territories of Reunion Island, French Polynesia and New Caledonia. To this end, in 2007 in accordance with Article 18 of Act 2006–436

---

[1] "Any area of intertidal or sub-tidal terrain, together with its overlying water and associated flora, fauna, historical and cultural features, which has been reserved by law or other effective means to protect part or all of the enclosed environment" (Kelleher 1999).

on National Parks, Marine Parks and Regional Nature Reserves, France created a specific institutional organisation, the *Agence des aires maritimes protégées*. To date, France managed to protect less than 1% of its exclusive economic zone (which is close to 11 million km², making her the second largest maritime country in the world). However, this percentage is far from reaching international commitments undertaken within the framework of the Convention on Biological Diversity, which aims to create a complete and coherent network of marine protected areas by 2012, representing 10% of marine ecosystems falling under national jurisdiction.

Marine protected areas address the challenges of marine biodiversity erosion and increasingly inefficient fisheries management (Chaboud and Cury 1998; Pauly et al. 2003; Hilborn et al. 2004). They have effectively become a laboratory for integrated coastal zone management (Davis 1998). When an MPA is established, there are economic, legal, geographical and social side-effects, along with visible, or less overt, territorial restructuring in which local, national and international NGOs all play a major role. Consequently, the study of MPAs requires a multidisciplinary approach: the legal and economic sciences are employed in this case to initially conceive the management systems, fit them into a legal and economic context at various levels, and finally evaluate the costs and advantages of MPAs in terms of resource conservation, and in terms of the economic development of the parties involved. In addition, bio-ecology has never before been used so extensively as the basis for conceiving MPAs and perfecting the indices for their evaluation and monitoring.

On the one hand, the effects of MPAs on ecological biodiversity have been widely studied (Russ 2002; Pelletier et al. 2005), while on the other, evaluations of any socio-economic benefits are very often partial and tend not to focus on the countries of the South (Oracion et al. 2005). Moreover, we should examine just how compatible the three major – and often competing – objectives of MPAs are: conservation of biodiversity, fisheries management and the promotion of nature tourism. As such, the performance of an MPA must be measured not only in biological terms, but also by taking social, economic and institutional considerations into account, in order to evaluate the pertinence and efficiency of MPAs as systems of governance.

Nevertheless, at this stage, the constraints on the implementation of MPAs as governance systems are often not fully appreciated by the researchers, users, managers, administrators and political decision-makers concerned. They are often not clearly explained to actors within civil society or to residents (Galletti 2006). The repercussions of such systems need to be evaluated, thus opening an important area of research which has been deferred for too long. While the consequences of MPAs on biological conservation appear positive (when they can be evaluated), those for resident populations and users are often disputed and somewhat ambivalent, making MPAs difficult to both justify and defend (Pelletier et al. 2005).

In this chapter, we will deal with marine protected areas from the viewpoints of the geographical, ecological, economic and legal disciplines since each of these disciplines offers different definitional approaches to issues regarding the governance of these marine – or semi-marine and coastal[2] – territories. In the first section, we intend to give an account of how these various scientific disciplines tackle the finer points of MPAs. We then show how the issues surrounding MPA governance require looking beyond these disciplinary approaches.

Marine Protected Area: A Specific Disciplinary Object for Geography, Ecology, Economics and Law

A Spatialised Approach to MPAs through Geography

While geography is concerned with spatial analysis, man-land relations, earth science and area studies, (Pattison 1964), only the first three concern protected areas. The landscape of an MPA differs from that of a protected terrestrial area in several ways: the presence of both a sea surface and a seabed; the absence of a permanent human component; and the lesser importance of topography in the structuring of landscape taxa[3]. In the field of coastal geography, the 40-year evolution which eventually brought the landscape concept to blend in with the geosystem concept (Richard 1989) is more complete (Corlay 1995; 1998). The geosystem therefore creates a bridge between the study of landscapes and spatial analysis.

A systematic examination of the coast makes it possible to view MPAs as a mechanism leading to the creation of various dynamic territories. These can refer for example to the space used by resident populations and their practices, and the effect this has on the resources of the MPA. They can refer to the geo-symbols[4] and representations of these populations vis-à-vis the resources, their habitat and usage. They can also refer to the regulatory dimension of zones implemented by the MPA management plan. Finally, they can refer to a territory encompassing the consequences of regulation, and the subsequent changes in use by populations residing in MPAs (David et al. 2006).

This type of territorial creation is inherent to any protected area, but due to the higher human density in coastal areas, and due to the importance of subsistence

---

2   In support of these analyses concerning the specifics of the marine, littoral and coastal environment, are elements from the scientific literature concerning MPAs, research conducted by the IRD on the protection of coral ecosystems in Oceania and the Indian Ocean, and elements of the research on marine areas in Madagascar supported by the trans-departmental incentive programme "Protected Areas" of the IRD, 2004–2005.

3   Indeed, knowledge of the detailed bathymetry of the shallow depths that characterise coastal MPAs remains very incomplete, due to a lack of appropriate tools for evaluation; therefore landscape taxa only include geomorphologic and bionomic information obtained through aerial or satellite remote sensing, along with ground verification.

4   J. Bonnemaison (1981) defines geo-symbols as places and itineraries appropriated by man over generations, which include culture.

and commercial fishing to the coastal economy, the creation of a protected area induces stronger reactions from local communities than it does in terrestrial areas. Geography then analyses the diachronic and synchronic dynamics between these spatial objects associated with the creation of any MPA. In addition to this are the links between the MPA and its surrounding territories, and the terrestrial territory of the resident populations within the MPA in particular, the space of exploitation that concentrates on the fringe of the MPA, and the new fishing space created by fishing aid obtained as compensation for the creation of the MPA. The geographer can appreciate the specificity of the MPAs according to the following points.

At the local level, the MPA is a territorial creation, found at the interface between coastal eco- and socio-systems. It partly modifies the direction of matter and information flows. For this reason, it constitutes a spatial-temporal discontinuity, comparable to a 'freeze' of space-time which equates to a 'freezing' of time concerning that space. This discontinuity takes on several forms. An MPA is a form of management and governance space that too often works like a closed system, having only very minimal links to the watershed and the local socio-economic environment, if only to minimise poaching or to protect any profit to be made from the environment. From the point of view of tourism, an MPA is an attractive space that sometimes generates at its fringe a concentration of hotels and sea diving clubs. From a fisherman's point of view however, an MPA is a prohibited zone that leads to the displacement of his fishing activity towards other places and species. Nonetheless it can also be an attractive space (edge effect). MPAs have complex effects on fishing grounds: exclusion in deferred fishing areas, but sometimes a reallocation within the space and according to the targeted species. They also imply a repositioning of fisheries within the coastal system of activities: activities involving the direct exploitation of marine resources within MPAs being partly or completely pushed away, to benefit tourism or non-exploitative activities. Managers networks are also being constituted as a response to a demand expressed by international conservation NGOs.

Compared to terrestrial protected areas, MPAs show a greater vulnerability vis-à-vis the local environment. In this regard, because they are less conscious of ecological concerns than hunters, fishermen are often opposed to MPAs. MPA sustainability depends on the integrated management of the coast on either side, including the consideration of watersheds, so as to reduce the terrigenous and pollutant flows generated by these areas (David et al. 2007).

A Natural Space for the Ecology

The Ecological Objectives of Conservation The Convention on Biological Diversity as well as the initiatives sponsored by major international NGOs such as the WWF or the IUCN, resulted in significant advances that led to the birth of MPAs. In addition to an increase in the number of protected spaces, the intention is also to increase the number of different habitats, paying special attention to endangered species and under-represented ecosystems, such as the open sea, which includes the problems of migratory species. The objective is also to take into account ecologically important ecosystems which deserve efficient protection and monitoring, such as seamounts or tropical and cold-water coral reefs. The challenges of conservation can imply the protection of essential habitats, such as spawning and hatching areas, by implementing temporal or geographical restrictions. They can also enable the maintenance of the functionalities of the ecosystem and the establishment of marine corridors between MPAs, to favour the resilience of ecosystems to climate change. These challenges are often met with a lack of knowledge about ecological dynamics, particularly in highly diverse ecosystems such as coral reefs. However, these environments are subject to special attention, not only in terms of protection and the implementation of management plans, but also in terms of research. These highly diverse ecosystems face considerable anthropogenic pressures, alongside the realisation that there has been a significant degradation of these environments on a global scale (we can mention for example the request for the Coral Reef of New Caledonia to be classified as a UNESCO World Heritage Site).

The Coral Ecosystem Example   The concept of 'reserve' corresponds closely to a traditional management measure used for centuries to protect coral ecosystems in the regions of South-East Asia or the Pacific Ocean, and which is increasingly being used in many other areas (Johannes 2002). Coral ecosystems are particularly exemplary of MPAs problems as they highlight bio-ecological issues. Gathering together various interconnected environments (e.g. sea grass beds, mangroves, reefs and channels), coral ecosystems constitute a network of habitats essential to the life cycle of species (i.e. reproduction, feeding, growth and refuge). This range of ecosystems also represents many potential fishing zones. This diversity of habitats explains the wide biodiversity of coral ecosystems. On a local scale, the high natural fragmentation of the habitat is due to the morphology of reef constructions. Unlike lagoons, their external slopes are more subject to the oceanic larval recruitment process and less exposed to anthropogenic pressures; they are also more interesting as far as following the impact of climate change is concerned. On a regional scale, the communities encountered from one island to another are clearly defined in spatial terms. The biodiversity of fish, plants and invertebrates depends on the geographic position of the island (the biodiversity gradient decreases eastward in the Pacific Ocean and westward in the Indian Ocean), on its type (open or closed atoll, atoll or high island), on its size, and on its degree of isolation. The natural fragmentation of coral ecosystems, which operates on a local and regional scale, is therefore one of the essential factors to be taken into consideration in designing MPAs and establishing MPA networks. The degree of protection afforded to the biological communities inside the reserves, and the degree of influence on adjacent areas, will depend on the size and spatial distribution of MPAs. Thus, the management plan implemented in 2004 for the maritime space of Moorea Island in French Polynesia, includes a network of eight MPAs, each delimited from the coast out to the barrier reef, preferably close to a channel, as they take into account the ecological criteria used to define the size and location of the areas to be protected.

Larval Dispersion and the Exchange of Nutritive Substances   Larval dispersion is an important phenomenon. In addition to the main outcome expected from an MPA, i.e. the restoration of reproductive stocks *inside* the deferred fishing area, one of the desired effects is the export of the biomass of the species for exploitation outside that area. The life cycle of the majority of marine species living in reef environments is divided into two distinct phases: the first one is a pelagic phase and concerns the eggs and/or larvae; the second one, relatively sedentary, concerns juveniles and adults. Larval dispersion explains the low rates of endemism and species extinction in marine ecosystems when compared to terrestrial environments. It also divides MPAs into two categories: those that export larvae, and those that receive them. In the first case, the local population is largely a result of self-recruitment. In the second case, it depends on the recruitment of larvae coming in from other populations (Shanks et al. 2003). Therefore the future of the MPA as an effective conservation tool requires that any coast to which it is linked through larval flow also be protected.

The exchanges of nutritive substances between adjacent ecosystems such as mangroves and coral reefs, and the interactions between pelagic and benthic zones (i.e. open and deep waters), or between coasts and coastal waters, must also be taken into consideration during the creation of MPAs, even if their management processes cannot prevent sediments, pollution or invading species to make their way into the protected area (Allison 1998; Simberloff 2000).

Ecological Implications of Temporal Variability   While space represents the main factor which structures coral ecosystem communities, hence the importance of spatial management techniques, time must also constitute a key parameter. It is necessary to study the link between this parameter and geographical scale and the biological processes concerned. Inter-annual variability is dictated by climatic phenomena on a global scale, while variability over shorter time periods can be explained by nycthemeral (or daily), lunar or seasonal cycles. In 1998, the massive episode of coral bleaching in the reefs of the Indian Ocean showed that MPAs in no way constitute protection against this type of threat. Vulnerability to bleaching constitutes an important criterion for the location of future MPAs, the emphasis being placed on stocking the most resilient reefs (to shield them from anthropogenic pressures). Migrations linked to the life cycles of species must also be taken into account[5], for the same reason as all the space-time-system biological interactions, so as to include them in the location and regulation of MPAs.

---

5   Reproductive migrations consist of species gathering in certain sites during reproduction periods; ontogenetic migrations correspond to the movement of cohorts (groups of individuals of the same age) during growth; trophic migrations correspond to the movement of individuals between two distinct habitats in order to feed.

Which Ecological Indicators Should Be Used to Follow and Evaluate MPAs? The establishment of 'zero points' and monitoring requires one to define indicators that take into account the expected effects MPAs have on the environment, from the viewpoint of management objectives, the capacity and response time of natural communities, and the functional characteristics of species (Adjeroud et al. 2005; Pelletier et al. 2005; Clua et al. 2005; Chabanet et al. 2005). Ecological indicators recommended for MPAs monitoring usually concern emblematic species, species targeted by fishing, as well as the biodiversity and global characteristics of the community and/or the quality of the habitat. They must be defined according to the management plan of the MPA, as well as its primary objectives, and the constraints of the organisation responsible for implementing such a plan. These objectives evolve over time[6] and imply new information about the biological systems and regular reassessment of regulations and their implementation.

A Specific Territory and Place of Activity for the Economy

Boersma and Parrish (1999) explain that economic objectives are preponderant in the creation of MPAs, due to the economic value of the ecosystems hosting them. In an attempt to estimate the monetary value of the environmental services provided by the main ecosystems found on earth, Costanza et al. (1997) have allocated to the coastal ecosystems (that are the most affected by the creation of MPAs) an average value of 4,052 \$/ha which, by comparison, is higher than that of tropical forests (969 \$/ha). Among the coastal marine environments, the highest values have been allocated to estuaries (22,000 \$/ha), sea grass beds (19,000 \$/ha) and reefs (6,000 \$/ha), with ecosystem services differing according to the environment concerned, such as recreational services for reefs, and nutriment recycling in estuaries and sea grass beds. MPAs can maintain or restore these environmental functions and therefore the economy to which they contribute. In a recent article, Martinez et al. (2007) confirmed the economic importance of coastal areas and oceans, in that they apparently represent between 60% and 70% of the total value of the world's ecosystems.

While MPAs are envisaged as promising for managing marine and coastal resources (Russ 2002), to what extent is the MPA more efficient than other forms of fishing regulations? Although, in the frame of the Ecosystem Approach to Fisheries, they are proposed as an alternative to conventional management methods, MPAs are not considered a panacea, but a tool which is essential to the sustainable use of resources (Cury and Miserey 2008). With regard to the economy, aside from the creation of wealth by ecosystems, major questions remain about the distribution of wealth and social justice. Since MPAs affect highly valued areas and resources, their creation causes intra- and inter-generational distribution effects,

---

6   Thus MPA managers are faced with the problem of assessing the impact of scuba diving or game fishing, which remain generally unappreciated, but which are more commonly practiced as efficient protection measures increase.

which consequently have an effect on their economic and social acceptability, and the requirement for a minimum equity criterion. The issue of the distribution of the economic impacts of MPAs over time is crucial: the opportunity costs that are borne when creating MPAs are immediate and certain, while the anticipated positive outcomes (economic and other advantages) are in the future and uncertain, especially when they are fundamentally linked to the maintenance or rehabilitation of environmental functions.

Finally, the MPA issue concerns the economics of institutions. The implementation of MPAs supposes that governance depends on certain conditions based on local particularism, as well as on models recommended by international environmental organisations. The success of MPAs depends largely on the quality of institutional arrangements and of collective action. In this context, economic enquiries are similar to legal enquiries: the quality of institutional constructions conditions the transaction costs during the creation of MPAs, and for their management, particularly if the MPAs rely on a consultative or participative model involving multiple stakeholders.

Moreover, tourism is experiencing a considerable boom in coastal areas. We have observed since 1990 (Hall 2001) a switch from mainly bathing tourism towards a form of tourism associating bathing activities with more sporting or adventurous activities (Chaboud et al. 2004). The establishment of an MPA results in the creation of an interface between an international market that considers the MPA to be a specific asset associated with a tourist destination, and environmental policies that try to limit the pressures of tourism on fragile ecosystems. In Madagascar for example, the development of tourism and ecotourism in particular, is considered to be a way of generating local revenues as compensation for the constraints imposed by conservation policies on more traditional uses. Some cases in the southwest of Madagascar show that a successful outcome depends on a set of economic and governance-related conditions that are rarely verified (Méral et al., this publication). We could at this point mention the governance of the international tourism industry in particular, which is not inclined towards the sharing of economic benefits equitably between local actors and operators upstream (Chaboud et al. 2004).

As far as economists are concerned, MPAs strongly crystallise the many challenges related to sustainability, and emphasise the links between the local and the international, along with the modes of governance applied to the territories concerned. Although the issues of economic evaluation and of sharing both the costs and the benefits seem crucial, they remain underdeveloped.

A Territory Governed by Law

Laws Related to the Study of MPAs   Although they are legal because they are created and managed within a legal framework (Froger and Galletti 2007), MPAs have only recently been studied by the law science discipline (Chaboud and Galletti 2007); perhaps because they are situated at the intersection of the law of the sea, coastal law (when it exists), environmental law and even economic law. The emergence of this new area of study is linked, on the one hand, to the growing status in the international law of the sea of the 'conservation' component, and to the obligations imposed on coastal states to regulate breaches in various maritime areas falling under their responsibility. On the other hand, it is linked to the interest which international environmental law (international and regional conventions) takes in marine areas that are either 'simple' or formed into networks, clusters or corridors (for more details on this notion, see Carrière et al., and Bonnin, this publication). Although environmental law came after the creation of the first protected areas, today it is one of their principle supports. Nor have MPAs developed independently of the bias of fishing regulations towards the preservation of protected areas (reservoirs of fish resources). Finally, MPAs reveal the status of modern law: they expose the existence of indigenous law and historical users of marine and coastal spaces. This concerns 'customary' or 'traditional' reserves or MPAs, as well as the related issues of the integration, opposition or recognition by 'modern' law, of pre-existing local rights concerning the coastal marine space.

The Role of the State and the Juxtaposition of Legal Competences On a national scale, marine ecosystems and MPAs are not spaces without rights or regulations. They contain legal systems such as the 'maritime public domain' that are different from those of the terrestrial 'public domain' or 'private domain of the state'. Marine ecosystems also accommodate many sectors (maritime traffic and trade, the tourism industry, industrial and local fishing, etc.) in which the state is strongly involved (via government departments or specialised institutions, amongst others), and where public and economic law is appealed to in a way which is different to that relating to terrestrial territories. Historical elements too lie behind the state's presence in maritime and coastal zones: control of the national maritime territory for law, public order and policing (with the involvement of departments such as the Home Affairs or Defence); state intervention in the fishing sector; the determining legal principle of state sovereignty over fish and mineral resources. One must not lose sight of these aspects when discussing the administrative and political systems of MPA management. The case of the marine and coastal protected area (MCPA) is even more particular in that it calls for the amalgamation of aspects of the law of the sea with other laws pertaining to the management of the terrestrial or coastal land. Institutions specialised in marine environments will find it difficult to manage the terrestrial space of an MCPA, and vice-versa. The difficulties experienced by island countries confronted with these issues are often given as an example in this regard.

The current rapid expansion of marine protected areas, particularly in the intertropical zone, is in line first of all with the historical increase in protected areas in terrestrial environments. However, the points of view of various disciplines towards MPAs have highlighted a number of issues that are due, among others, to the significance of economic stakes and claims to access resources and spaces. For example, the commercial exploitation of the living resources of the sea, which has no equivalent in the terrestrial environment, will from now on be required to coexist with the interests of tourism or conservation. When considering the various scientific disciplines involved as a whole, the issue of governance turns out to be central, although it is advisable to know whether the disciplinary approach is still sufficient to answer it.

Towards a Multidisciplinary Approach of MPAs and of their Governance Mechanisms

What we seek to identify is neither rupture nor continuity in the scientific study of MPAs but, rather, the shift or perhaps even the transformation of MPAs as perceived from the viewpoint of each discipline, towards a new multidisciplinary entity. Moreover, the combination of their particularities has consequences for the study and conception of those MPA governance plans that are less mono-disciplinary but more experimental and receptive to disciplines other than law and economics, both usually concerned with the administration of territories and public choices. There is a tendency among public and private organisations to create MPA governance that relies on all the discipline-related information that could be collected.

From MPAs as mono-disciplinary units to MPAs as Multidisciplinary units

MPAs are complex units, and any reading of them from the strict viewpoint of certain disciplines, only tackles a portion of the sets and relations defining them. Understanding the structure of an MPA system requires knowledge from various disciplines, if the MPA is to be understood in all its diversity.

Experts interested in MPAs are few, irrespective of the country being considered. The increase in the number of MPAs in the inter-tropical zone has subsequently generated a growth in the demand for multidisciplinary studies, while the supply of expertise has not improved[7], even if some progress is perceptible[8]. Gathering a team of experts from different disciplines is a rare achievement. Too often research

---

7   The low scientific supply from the countries of the South can be explained by the small number of researchers specialising in ecosystems and the even smaller number of researchers specialised in coastal socio-systems. Generally, social science departments show little interest in coastal environments, and even less in marine environments. Students trained in rural or urban studies prefer to invest their skills in urban or rural studies rather than in coastal or marine research sites.

8   In East Africa, a study conducted within the Western Indian Ocean Marine Science Association aims at federating researchers working on the coasts of nine countries, so as to develop multidisciplinary degree courses that, together with oceanographers, will train 'coastal and marine' generalists and researchers in social sciences who will have a good knowledge on the coastal environment.

teams are limited to bringing an expert from the biological sciences together with an expert from the social sciences, thereby amalgamating disciplines as diverse in their issues and methods as anthropology, law, economics, history, geography, political science and sociology. This situation forces researchers belonging to a particular discipline to turn their attention to the fields of related disciplines, and even those of more thematically remote disciplines, that could make an indispensable contribution to understanding the MPA system. The mixture of skills used from several disciplines, not always well assimilated, tends to generate a multidisciplinary approach that could hardly qualify as 'a hybrid science of MPAs'.

The combination of the specific characteristics of MPAs almost ineluctably determines a multidisciplinary approach, if not a transversal one. Indeed, the geographical or economic particularities of MPAs have legal implications that should enable the manager to differentiate between marine and terrestrial protected areas; as far as their functional management system and their administration plan are concerned.

Thus, as an open maritime space, an MPA implies control and monitoring difficulties that cannot be compared to those of protected areas on land, and gives cause for conflict between management institutions and economic operators. MPAs made up of areas situated along the coastal fringe complicate coastal development policies and the legal relationships between elected people, local actors and tourism operators. Moreover, areas that can be transformed into MPAs are often subject to amplified anthropogenic pressure, as a result of economic actors exploiting the coastal resources. In this regard, the de-/centralised public administrations that control these activities and these human flows, tend to deploy the legal arsenal intended to guarantee the efficient regulation of environmental infringements and economic transactions. Traditional controls over the maritime domain, often military, have in fact never completely disappeared, and are revived in moments of conflicts of interest between economic actors; conflicts which the state intends to regulate and solve. The simple fact that the states of the South face a serious shortfall in financial and logistical resources does not change their attempt to control space, even if it often remains purely theoretical. Ultimately, MPAs inevitably become a separate category of space to be protected (Chaboud and Galletti 2007).

The Specificity of MPAs and the Consequences for Governance

Researchers and decision-makers have been forced to treat MPAs and their governance systems as rhizomes, at the junction between nature and society.

In this case, the geographic input to MPA governance is progressive. The temporal dimension takes on great importance when clarifying the human-environment interactions and related governance processes: protected areas were initially conceived on an island model that progressively transformed into a reticular model, based on ecological corridors (Carrière et al. and Bonnin, this publication). Favouring, as it does, accessibility to biodiversity and its touristic valorisation, this evolution has implications for terrestrial protected areas (Grenier 2003). At sea, the reticular model in fact became essential to biologists due to the aquatic environment's inherent 'permeability' to larvae and juveniles. On the other hand, as regards governance, MPAs have been designed according to the terrestrial protected area model as spatial discontinuities (Gay 2003; David 2003). On land, the recent generalisation of the buffer zone concept, as introduced by UNESCO in the 1970s in its biosphere reserves, reduces the discontinuities between an area which is protected and an area which is not. On the contrary, in the case of MPAs, one notes that in the marine environment, there is an emphasis on discontinuity, since the fringes of the protected areas are subject of increased anthropogenic pressure on resources such as augmentation in fishing activity and the diversification of tourism activities (e.g. the development of snorkelling). In order to limit the emphasis of the discontinuity in the reef environment, it is generally proposed that a new marine space be created further offshore, via the implementation of fish concentration systems, so as to transfer fishing activity beyond the reef area (David 1998).

Nonetheless, when the monetary value of the exploited resources is high, it is indispensable to associate the regulation of the fishery pressure on the peripheral maritime areas with the creation of MPAs in order to avoid the overfishing of protected species. This measure must be complemented on land with the creation of activities that can generate revenues, generally considered by the local communities as a "fair reward" for their involvement in the management of the MPA.

The example of the marine park of Mohéli in the Comoros shows that, when the governance of the protected area is efficient, managers are sometimes requested to extend their intervention to the entire terrestrial region of the resident communities affected by the MPA (David et al. 2003). The spatial integration of land and sea within the same MPA is a new approach which is sometimes implemented at the *national* level, to promote the 'protected area' as a tourist product as complete as possible. On the other hand, at the *international* level, biodiversity protection is still influenced by the essential dichotomy between marine and terrestrial environments, as illustrated by the results of the so-called ecoregional approaches promoted by the WWF. The emphasis here is increasingly placed on the identification of centres of biodiversity and the spatial relations that exist between them.

In this context, taking into account the ecological connectivity of the reefs today leads to the inclusion of a *regional* dimension in the governance of MPAs. Thus, in the Indian Ocean, the creation of a regional network of MPAs is currently the subject of a programme guided by the WWF under the aegis of the Indian Ocean Commission, with financing from the French Global Environment Facility. This regional dimension is intended to reduce the discontinuities created by MPAs in relation to the surrounding environment. It also happens that in case of potential conflicts between states over coastal or marine resources (e.g. oil or fish), MPAs are used in a regional perspective as indicators of geographical discontinuities, or even as factors for the increase of these discontinuities when MPAs serve as politically neutral buffer zones, in which case the management of the border territory is passed on to an international NGO.

Regarding the organisation of a governance system, the law takes on a privileged role due to the legal functions and dimensions of the MPA administrative system. As such we should mention the production mode of the rules applied in MPAs, the conditions for the application of these rules, the evaluation of the daily functioning of the administration, management, financing, control, sanction, negotiation and regulation of the crises.

The field of economics, which is also concerned with the study of MPAs, can collaborate with the field of law. The institutional change provoked by the creation of MPAs produces multiple and sometimes counter-intuitive effects, making it impossible to carry out simplistic or naïve analyses. Resorting to several disciplines then becomes an advantage. Thus, the economic appraisal shows that the participative model which underlies the delegation of management which is often implicit in promoting good governance, results in transaction costs that can reduce the efficiency of MPAs in pursuing their conservation and local sustainable development objectives. The 'proliferation of institutions' represents a threat to the establishment of MPAs in the South, and institutional rent seeking is sometimes counterproductive (Baghwati 1982), considering the objectives attributed to MPAs. As far as the legal field is concerned, the difficulties of the participative model could mean a return to state interventionism as far as conception and methodologies are concerned. State interventionism can reappear, either in the case of disagreement or in the stalling consensus between non-governmental actors concerned with MPAs, or in case of discord between administrative interventions and local civil practices (the way of life, consumption and exploitation of inhabitants, and historical actors on the natural environments and resources). Interventionism can also reappear when an MPA sponsor retires and the pressure it exercised on the administrative services eases, to the benefit of a more local management.

Beyond the networks of marine and/or coastal sites classically managed by the state, in the case of a participative governance (Féral 2007), some areas are co-managed with local communities and other stakeholders, while private protected areas are managed by their owners. These more territorialised plans are part of the movement in favour of bottom-up conservation. They are the outcome of the continued search for successful MPA governance, co-management being currently a trend as far as environmental governance is concerned. However, transformations in MPA governance become apparent in specific cases: sometimes co-management becomes the default solution when the state is unable to manage MPAs. In such cases the only cost to the state is to legalise its implementation; the political cost is higher in that it involves, for the state, a certain loss of sovereignty by relinquishing its centralised interventionist prerogatives. Sometimes co-management is a success ascribed to both parties.

The state's reappearance is by its very nature political. Since the 1972 Stockholm and 1992 Rio de Janeiro Summits, as well as the 2003 Durban World Parks Congress, environmental protection – and therefore the creation of MPAs – has become a major international challenge.

This development is particularly visible in some island states of the Indian Ocean or Oceania. The regime of President René in the Seychelles used it, at the end of the 1970s, to build a respectable image for the country. The political advantage gained from taking the international stage ended in significant economic costs for the country, since the creation and management of the MPAs were entirely paid for by the Seychelles government. Since then, the situation has evolved considerably. Powerful international NGOs, North American NGOs in particular, have been financing the majority of – if not all – the implementation of protected areas. The costs to the recipient state are therefore modest compared to the political benefits yielded by the operation.

Countries from Oceania with limited economic resources, such as the Federated states of Micronesia and the Marshall Islands, have embarked on this type of operation with the intention of taking advantage of both political and economic benefits. A parallel can be drawn with the Malagasy presidential commitment in 2003 in Durban, concerning the expansion of protected areas, marine in particular, which pulled together sponsors, NGOs and state institutions.

However, MPA implementation in Oceania and the Indian Ocean is still constrained by difficulties and a lack of definition. In addition, the financial and social costs of conservation, already high for terrestrial protected areas, remain a problem for future MPAs. In Madagascar, as in the poor countries of the Indian Ocean, due to the multiplication of protected areas, the state could end up without the means to carry out its prerogatives as both MPA creator and manager, leaving the many international NGOs or local associations a clear field to act. This can be explained by the decrease in the state's effective intervention capacity due to drastic cuts in public expenditure, while environmental protection is increasingly included in the conditionality of public development aid.

In reality, despite the massive withdrawal of the state when it comes to either financing the MPA operation (particularly via the creation of trust funds managed by international NGOs and in which a state is a partner, among others, see Méral et al., this publication), or negotiating with local communities, the state always remains present. Indeed it cannot be avoided as far as the administrative aspect, MPA registration and the associated legalities are concerned (i.e. integrated management of coastal areas, decentralisations, association law, resorting to agencies and the legal framework of the fishing or tourism sectors). Even in the case of Madagascar or the Comoros, where the administration is falling apart in such a way that sponsors are taking the MPA creation process into their own hands (by looking after the financing and current management of the MPA when it cannot be supported by the state), sponsors still cannot break away from the state which is indispensable in order to ratify the legal status, the zoning or the policing of the MPA. The steps taken by sponsors towards the state are permanent. Once the legal framework is completed, sponsors often ensure its management, or promote its delegation, with an NGO or agency.

Special Governance between Authorities and Private Organisations

The MPA sector is becoming specialised due to the efforts (in terms of time, design, means and projects) made by public decision-makers and dedicated institutions. The differences between MPAs from various states are becoming less marked while their experiences are becoming increasingly similar. Their failures and oppositions too are often similar. A culture of MPA and MCPA managers could just as well be created, with the appearance of a body of civil servants or private experts specialised in this domain.

In the majority of cases, we find that an effort has been made to clarify the legal situation concerning MPAs, and that the authorities in charge of MPAs are becoming aware of the new influence of administrative and political decentralisation, as well as local authorities. Finally, we observe attempts by state administrations to better align their conservation action with the interests of territorial organisations (e.g. territorialised structures and groups of local stakeholders). These attempts are two-fold. On the one hand, the state tries to create *legal* governance for MPAs, which cannot always be autonomous in developing countries; and on the other hand, this legal governance should rely on a decentralisation process (where the decentralised authorities become environmentally competent). The state sometimes also wants to legitimise the 'de facto' practices of pre-MPA actors who hold certain powers, in which case the administrations adopt unchanged or updated rules on local access, and use the efficient self-monitoring ability of local individuals and groups on protected areas. The focus is on regulations peculiar to local actors. Such regulations remained little known for a long time and were often considered archaic. The state now wants to bring it into law in order to overcome the inefficiency of modern instruments. The renewed attention given to systems of sanctions expanded to include the protection of natural resources, by reusing existing customs and sometimes reinterpreting them, is a case in point.

Above MPA management, the central state, at the highest institutional level, can adopt two positions. On the one hand, it can co-ordinate the management and legal actors involved in a given maritime area, which will paralyse it if it cannot manage to juridically organise this institutional complexity in competition with its own. On the other hand, it can revert to the (opposite) centralised unilateral formula, aimed at determining the MPA perimeter as a space distinct from ordinary areas, a special area where the rules of common law are excluded to the benefit of more restrictive access, harvesting, displacement and development methods. Through this, the state includes the additional territory into a new grid pattern, delineating MPAs which state agents can dominate. The management of this area can be left either to a public institution (with reduced material and human resources) or to a private establishment created for that purpose. This management institution can be linked to the state or to private sponsors or NGOs, in proportion to the quantum of private funds and/or exogenous funds invested to ensure this management function.

Conclusion

The recent increase in the number of marine protected areas does not yet include specifically dedicated autonomous public policies. On the one hand, MPA zoning which benefits from special regulatory and administrative policing, is more the concern of a programme or a simple project, than a carefully considered public policy. On the other hand, MPAs are often included in an all-encompassing national policy (e.g. environmental protection, fishing or forest management, coastal development or the integrated management of coastal areas) and are only a particular element of it. It is important to take MPAs out of the policy context that sometimes obscures their analysis more than it helps it, by keeping in mind that whereas MPAs, as conservation tools, are unequivocally part of the *conservation of natural resources*, they can prejudice any improvement in the livelihoods of the most disadvantaged individuals and social groups. And yet, they are part of sustainable development that advocates the *pursuit of poverty reduction and best distribution of wealth between beneficiaries of development* (Chaboud 2006).

References

Adjeroud M., Chancerelle Y., Schrimm M., Perez T., Lecchini D., Galzin R., Salvat B., 2005 – Detecting the effects of natural disturbance on coral assemblages in French Polynesia: a decade survey at multiple scales. *Aquatic Living Resources*, 18: 111–123.

Allison, G. W., 1998 – Marine reserves are necessary but not sufficient for marine conservation. *Ecological Applications*, 2: 79–92.

Baghwati J. N., 1982 – Directly unproductive profit-seeking activities. *Journal of Political Economy*, 90: 988–1003.

Boersma P. D., Parrish J. K., 1999 – Limiting abuse: marine protected areas, a limited solution. *Ecological Economics*, 31: 287–304.

Bonnemaison J., 1981 – Voyage autour du territoire. *L'Espace Géographique*, 4: 249–262.

Chabanet P., Adjeroud M., Andrefouët S., Bozec Y. M., Ferraris J., Garcia-Charton J., Shrimm M., 2005 – Human-induced physical disturbances and indicators on coral reef habitats: a hierarchical approach. *Aquatic Living Resources*, 18: 215–230.

Chaboud C., 2006 – Gérer et valoriser les ressources marines pour lutter contre la pauvreté. *Études Rurales*, 178: 197–212.

Chaboud C., Cury P., 1998 – Ressources et biodiversité marines. *Natures, Sciences Sociétés*, 6 (1): 20–25.

Chaboud C., Galletti F., 2007 – Les aires marines protégées. Une catégorie particulière de territoires pour le droit et l'économie ? *Mondes en Développement*, 35 (138): 27–42.

Chaboud C., Méral P., Adrianambinina D., 2004 – L'écotourisme comme nouveau mode de valorisation de l'environnement: diversité et stratégie des acteurs à Madagascar. *Mondes en Développement*, 32 (1): 11–32.

Chaboud C., Froger G., Méral P. (eds.), 2007 – *Madagascar face aux enjeux du développement durable. Des politiques gouvernementales à l'action collective locale.* Paris, Karthala, p. 308.

Clua E., Beliaeff B., Chauvet C., David G., Ferraris J., Kronen M., Kulbicki M., Labrosse P., Léopold M., Letourneur Y., Pelletier D., Thébaud O., Leopold M., 2005 – Towards a multidisciplinary indicator dashboard for coral reef fisheries management. *Aquatic Living Resources*, 18: 199–213.

Corlay J. P., 1995 – Géographie sociale, géographie du littoral. *Norois*, 165: 247–265.

Corlay J. P., 1998 – "Facteurs et cycles d'occupation des littoraux". *In* Miossec A. (ed.), *Géographie humaine des littoraux maritimes*. Paris, CNED-SEDES: 97–170.

Costanza R., d'Arge R., de Groot R., Farber S, Grasso M., Hannon B., Limburg K., Naeem S., O'Neill R. V., Paruelo J., Raskin R. G., Sutto P., van den Belt M., 1997 – The value of the world's ecosystem services and natural capital. *Nature*, 387: 425–259.

Cury P., Miserey Y., 2008 – *Une mer sans poissons*. Paris, Calmann-Lévy, p. 279.

David G., 1998 – "Les aires protégées, laboratoires de la gestion intégrée des zones côtières: l'exemple des pays membres de la Commission de l'océan Indien". *In* IIe rencontres *Dynamiques sociales et environnement*, Bordeaux 9–11 September 1998, UMR-Regards CNRS/Orstom, vol. 2: 343–360.

David G., 2003 – "Les aires protégées littorales de la zone de la Commission de l'océan Indien". *In* Lebigre J. M., Decoudras P. M. (eds.), *Les aires protégées insulaires et littorales tropicales.* Bordeaux, University of Bordeaux 3, CRET, coll. Îles et archipels, 32: 55–72.

David G., Lo H., Soule M., 2003 – "Le parc marin de Mohéli (Comores), de la protection des tortues à la gestion de l'espace insulaire". *In* Lebigre J. M., Decoudras P. M. (eds.), *Les aires protégées insulaires et littorales tropicales.* Bordeaux, University of Bordeaux 3, CRET, coll. Îles et archipels, 32: 121–135.

David G., Mirault E., Quod J. P., Thomassin A., 2006 – "Les concordances territoriales au cœur de la gestion intégrée des zones côtières : l'exemple de la Réunion". *In* Colloque *Interactions nature-société, analyse et modèles*, La Baule, 3–6 May 2006, http://letg.univ-nantes.fr/colloque/actes.htm.

David G., Antona M., Botta A., Daré W., Denis J., Durieux L., Lointier M., Mirault E., Thomassin A., 2007 – *La gestion intégrée du littoral récifal de la Réunion : de la connaissance scientifique à l'action publique, jeux d'échelles et jeux d'acteurs. Prospective du littoral, prospective pour le littoral, un littoral pour les générations futures.* Paris, La Documentation Française/Ministère de l'Écologie et du Développement durable.

Féral F., 2007 – L'administration des aires marines protégées en Afrique de l'Ouest. *Mondes en Développement,* 35 (138): 43–60.

Froger G., Galletti F. (eds.), 2007 – Regards croisés sur les aires protégées marines et terrestres. *Mondes en Développement,* Special issue, 35 (138), p. 138.

Galletti F., 2006 – "Quelle(s) gouvernance(s) pour le développement durable face à la mondialisation. Le cas de Madagascar. Introduction à la Partie Troisième". *In* Froger G. (ed.), *La mondialisation contre le développement durable ?* Brussels, Peter Lang, Presses interuniversitaires européennes: 218–233.

Galletti F., 2007, – "La gestion durable de la biodiversité dans un pays en développement". *In* Méral P., Froger G., Chaboud C. (eds.), *Madagascar face aux enjeux du développement durable. Des politiques environnementales à l'action collective locale.* Paris, Karthala: 81–105.

Gay J. C., 2003 – "Discontinuités et aires protégées". *In* Lebigre J. M., Decoudras P. M. (eds.), *Les aires protégées insulaires et littorales tropicales.* Bordeaux, University of Bordeaux 3, CRET, coll. Îles et archipels, 32: 17–27.

Grenier C., 2003 – "Discontinuité et accessibilité des aires protégées: du modèle insulaire au modèle réticulaire". *In* Lebigre J. M., Decoudras P. M. (eds.), *Les aires protégées insulaires et littorales tropicales,* Bordeaux. University of Bordeaux 3, CRET, coll. Îles et archipels, 32: 29–42.

Hall C. M., 2001 – Trends in ocean and coastal tourism: the end of the last frontier? *Ocean and Coastal Management,* 44: 601–618.

Hilborn R., Stokes K., Maguire J. J., Smith T., Botsford L.W., Mangel M., Orensanz J., Parma A., Rice J., Bell J., Cochrane K. L., Garcia S., Hall S. J., Kirkwood G. P., Sainsbury K., Stefansson G., Walters C., 2004 – When can marine reserves improve fisheries management? *Ocean and Coastal Management,* 47: 197–205.

Johannes R. E., 2002 – The renaissance of community bases marine resource management in Oceania. *Annual Review of Ecology and Systematics,* 33: 317–340.

Kelleher G., Bleakley C., Wells S., 1995 – *A global representative system of Mmarine protected areas.* Washington, The World Bank/IUCN, 4 vol.

Martinez M. L., Intralawan A., Vasquez G., Perez-maqueo O., Sutton P., Landgrave R., 2007 – The coasts of our world: ecological, economic and social importance. *Ecological Economics,* 63 (2–3): 254–272.

Oracion E. G., Miller M. L. Christie P., 2005 – Marine protected areas for whom? Fisheries, tourism and solidarity in a Philippine community. *Ocean and Coastal Management,* 48 (3–6): 393–410.

Pattison W. D., 1964 – The Four Traditions of Geography. *Journal of Geography,* 63(5): 211–216.

Pauly D., Alder J., Bennett E., Christensen V., Tyedmers P., Watson R., 2003 – The future for fisheries. *Science,* 302: 1359–1361.

Pelletier D., García-Charton J. A., Ferraris J., David G., Thébaud O., Letourneur Y., Claudet J., Amand M., Kulbicki M., Galzin R., 2005 – Designing indicators for evaluating the effects of marine protected areas on coral reef ecosystems: a multidisciplinary standpoint. *Aquatic Living Resources*, 18: 15–33.

Richard J.-F., 1989 – *Le paysage, un nouveau langage pour l'étude des milieux tropicaux*. Paris, Orstom, coll. Initiations-doc tech., p. 210.

Russ G. R., 2002 – "Yet another review of marine reserves as reef fisheries management tool". *In* Sale P. F. (ed.), *Coral reef fishes. Dynamics and diversity in a complex ecosystem*. San Diego, Academic press: 421–443.

Salvat B., Haapkyla J, Schrimm M., 2002 – *Coral reef protected areas in international instruments*. Perpignan, EPHE, p. 196.

Shanks A. L., Grantham B. A., Carr M. H., 2003 – Propagule dispersal distance and the size and spacing of marine reserves. *Ecological Applications*, 13 (1): S159–S169.

Silva M. E., Gately E. M., Desilvestre I., 1986 – A bibliographic listing of coastal and marine protected areas: a global survey. *Woods Hole Oceanog. Inst. Tech. Rept*. WHOI: 86–11.

Simberloff D., 2000 – No reserve is an island: marine reserves and indigenous species. *Bulletin of Marine Science*, 2: 567–580.

PART II
New Tools?

With the redefinition of protected areas through sustainable development, new tools are being developed and implemented: concepts of corridor and networks in conservation biology, geopolitical networks, financing mechanisms and legal regulations. The second part of this publication is interested in the ways in which these new tools are disseminated, and seeks to evaluate their efficiency and coherence in terms of political viability.

One can legitimately question the balance between ecological and social dynamics, or the degree of compatibility between theoretical conservation models and 'traditional' natural resource management. The nature and evolution of scientific, technical or local knowledge produced and mobilised to justify and organise protected areas, represents a scientific as well as public policy challenge. This can be observed in the development and mobilisation of new conservation tools, in the ecological and social domains.

The biological models of conservation currently question the efficiency of protecting isolated and well-delimited territories. The corridor concept advocated in this regard is presented as a network open to outside influences, so as to maintain different types of habitats as well as the exchange of genes between individuals from differentiated landscapes. The concepts of ecological network, ecotone, corridor and bio-geographical region are put forward to highlight the importance of connectivity through horizontal relations and exchanges. However, the use of 'ecological corridor' seems largely metaphorical when used in development and institutional networking projects.

In response to the appeal launched by its president, the Malagasy government is trying to triple the surface area of conservation sites. This fits into the global trend of protected area expansion. The concept of conservation corridor serves this dynamic by placing almost all the country's forests into a network, i.e. the forest relics that are quite appropriately and geographically part of some sort of corridor. Madagascar has also included poverty alleviation into the objectives of environmental policy. However, with the ambiguity of the concepts used and the sensitivity underlying the combination of economic and ecological approaches, the nascent economic networks (in the form of embryonic developments on a regional scale) are not linked – and some are even opposed – to ecological 'corridor' objectives. Stéphanie Carrière, Philippe Méral, Fano Andriamahefazafy and Dominique Hervé bring to light the way in which the concept of ecological corridor, contrary to the official discourse, is not used in the planning and management of protected areas. They show that a bio-focused top-down approach does not guarantee success for conservation and sustainable development, unlike the more promising approach based on the definition and identification of socio-ecological corridors by local populations.

Marie Bonnin examines the issue of ecological network (closely related to that of corridor) from a legal perspective, where it appears as a new structuring element in conservation policies. Indeed, the entire world of protected areas is affected by internationalisation, through the constitution of ecological networks acting as natural infrastructure, through the constitution of institutional networks of conservation management (via the exchange of experiences as well as transboundary co-operation), and through financing. There has been a notable increase in the number of political and legal texts requiring the authorities to establish ecological networks. In this respect, networking seems to concern mainly management institutions, with a view to mutualising infrastructures and skills. The evolution of institutional mechanisms necessitates placing into perspective the role of protected areas which, depending on the geographic scale or their transboundary character, can also act as corridors by enabling the maintenance of biological interconnections. Considering sustainable development could have the effect of multiplying complementarities between conservation areas and territories allowing human activities, towards a potential 'spatial dilution' of conservation into a more global land planning policy; in which case the role of peripheries or buffer zones is crucial to reduce discontinuities. However, protected areas integrated into ecological networks can become experimental zones of sustainable development, to the detriment of nature protection objectives.

In addition to corridors as biological and legal planning tools, innovative financial tools are also being developed. These tools resort to market solutions and to the negotiation of development rights. Indeed, protected area financing over the last few years experienced major disruptions, particularly in developing countries with high biodiversity. Philippe Méral, Géraldine Froger, Fano Andriamahefazafy and Ando Rabearisoa use the case of Madagascar to show how the current conservation policy, with the expansion of protected areas to attract more investments and their new role as environmental service suppliers, goes hand in hand with the development of new financial instruments. The rhetoric of the fight against climate change applied to protected areas transforms these into potential carbon sinks. It makes them part of 'carbon market' systems in which new and powerful investors intervene. These conservation financial markets are then delinked from biophysical phenomena in complex ways, and, through the complexity of their setup and the importance of capital outlays, are hardly adapted to the capacities of local actors for negotiation and appropriation.

Chapter 3

Corridors: Compulsory Passages?
The Malagasy Example

Stéphanie M. Carrière, Dominique Hervé,
Fano Andriamahefazafy and Philippe Méral

Since the 5th World Parks Congress held in Durban in 2003, the maintenance or restoration of corridors with a view to improving connectivity, has become a fundamental element of new conservation policies. The objective of networking protected areas and maintaining or rehabilitating corridors is to overcome the drawbacks of former conservation strategies, based as they were on protecting one isolated area from another, and to avoid the effects of ecosystem fragmentation leading to the loss of biodiversity. In order to assess whether this new tendency constitutes a rupture or a continuity in traditional policies used for establishing protected areas, we need to explore the meaning of 'corridor' and the pertinence of its application in the field of conservation. Indeed, the notion of 'corridor' is not peculiar to scientists or conservation actors; it is part of the common discourse. It has a multiplicity of meanings, and its use has spread throughout many disciplines, going back to the 1990s. Whereas it is certainly better known in the fields of conservation and ecology, the notion of 'corridor' is similarly employed in relation to the issues of contemporary economic and urban studies, land use planning, and even to the flow of goods, people and information.

Irrespective of the scale and disciplinary context, and despite its multiple meanings, a corridor is always defined by its elongated shape and its function as a conduit or obstacle to the flow of matter and information. But is this enough to make the step from an innovative theoretical concept to a working process for biodiversity conservation? We propose at first to deconstruct the concept of 'corridor', i.e. to analyse its origins and scientific interpretations within the disciplines of the life and social sciences. We then use this analysis to explain the confusions, the ambiguities and the controversies that the use of the term 'corridor' invokes, when used in the field of conservation. Finally, we use the example of the spread of conservation corridors in Madagascar to illustrate the results of our analysis. In this regard, the Malagasy environmental approach has adopted a conservation policy centred on corridors with a view to increasing the extent of protected areas. A clarification of the anticipated environmental and economic impacts of the corridor model is welcome in this country of widespread poverty and high endemism. Considering what is at stake in terms of sustainable development, the functions of a corridor must be clearly defined, especially when the goal is not only to create protected areas for the conservation of biodiversity, but also to contribute to the alleviation of poverty[1]. Indeed, the impact the creation of protected areas has on local populations is not well known, while the risks are far from negligible.

Corridors across Disciplines

The term 'corridor' initially comes from the field of conservation biology; however, it has been used in domains as varied as land use planning, and development economics[2].

From Game Reserves to Conservation Corridors: The Ecological History of the Concept

Corridors have a long history. At the beginning of the 20th century, they were first used to basically establish and maintain fauna in game reserves (Harris and Sheck 1991). Only later did corridors become a subject for study by scientists and a conservation tool for managers, culminating in a new science documented in a publication entitled *Corridor ecology: the science and practice of linking landscapes for biodiversity conservation* (Hilty et al. 2006).

The term 'corridor' was originally used by the first landscape ecologists in the 1940s (Forman and Godron 1986), particularly in relation to watercourses (stream corridors). A structural definition of the term linked to the elongated shape of corridors, hedges, streams, etc. then appeared. Only later did Forman and Godron (1981; 1986) introduce the matrix-patch-corridor concept which they applied to the landscape structures seen in aerial photographs and satellite images in order to describe and to analyse them. In this case, the 'matrix' is the dominant landscape element which is the most connected, while the 'patch' is a non-linear area and the 'corridor' is a linear entity (Figure 3.1). Significant vocabulary and literature describe the structure, origin, objectives and functions of corridors within this paradigm (Burel and Baudry 1999).

The corridor concept in relation to the biodiversity conservation appeared more recently, stemming from the island biogeography theory of McArthur and Wilson (1967) and from the meta-population theory (Levins 1969; McCullough 1996; Hanski and Gilpin 1997).

These two theoretical corpora form the basis of conservation biology, which advocates the use of corridors to improve the flows of animal or vegetal individuals and species. Thanks to the dynamic equilibrium theory (McArthur

---

1   Conservation appears as a contribution to the development of Madagascar in the *poverty reduction strategy paper*.

2   The international and legal dimensions of networks and corridors are tackled in the next chapter.

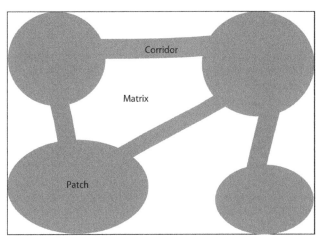

Figure 3.1   Spatial illustration of the 'Matrix-Patch-Corridor' model

and Wilson 1967), it is possible to predict the number of species present on an island, given its surface area and the distance to the closest continent representing a source of individuals (Blondel 1995). The global assumption underlying this theory sets out that species diversity on an island results directly from two dynamic processes: the colonisation rate of individuals and the extinction rate of populations. Consequently, the number of species can increase when the island is large and close to the mainland source (McArthur and Wilson 1967). This was the first theory on the influence of spatial organisation on ecological processes. Whereas this theory provoked many reactions and controversies, it also initiated much research.

From the 1980s onwards, the island model gave way to the meta-population concept, as set out by Levins (1970). It was on this research that the effects of habitat fragmentation on populations were based. A meta-population consists of small populations that become extinct and leave vacant habitats that are re-colonised locally. Furthermore, the permanence of a meta-population is only possible if the average extinction rate is less than the colonisation rate. Individuals who scatter can colonise vacant sites, and occupied sites can become vacant following local extinctions. These sites are in turn colonised by disperser individuals.

Many animal communities reflect characteristics that are accurately represented by this theory, or by theories derived from this one: the model of Boorman and Levitt (1973), the source-sink model[3] of Pulliam (1988) and Blondel et al. (1992). Local extinction processes can be dependent on the structure and dynamic of the landscape. As such, the isolation, size and shape of patches of habitat can influence colonisation and extinction rates. For example, the smaller a sub-population is, the

---

3   In this model, the meta-population consists of patches in which the growth rate is positive for certain (source individuals) and negative for others (sink individuals) (Pulliam 1988).

more likely it is to disappear in the face of demographic probability. Moreover, the size of sub-populations correlates with the size of their habitat, e.g. a small forest grove. The more groves there are and the closer these are to one another, the more the probability of extinction decreases, since the likelihood of immigrants arriving in each grove increases.

These theories underpin the work of conservation biologists. What is the potential role of corridors in the operation of the island model and the meta-population theory? The existence of biological corridors (forests, hedges and rivers) enabling the flow of disperser individuals between sub-populations, would theoretically favour the maintenance of meta-populations and therefore of the species in the long run. Indeed, individuals in certain species are reluctant to disperse into an environment which is not their own (in order to reproduce or feed) or which is not favourable to their survival (predation). The bridges that join similar ecosystems or sites are called 'corridors'. Their efficiency can be measured in terms of the flow of disperser animals, and therefore of genes, for the specific colonisation of small seasonally-interconnected populations (Fahrig and Merriam 1985, for the micro-mammals of the Ottawa region). These authors have shown that patches are re-colonised each spring, and that animals prefer to travel along the hedges found between groves. An increase in the number of these corridors increases the connectivity between patches, which then increases the survival time of the meta-population.

Corridors have been given a role to play in the conservation of forest ecosystems, particularly by overcoming the potential effects of their fragmentation, the resulting isolation of their animal and plant populations, and even their extinction. Managers and conservationists whose responsibility it is to protect species, will try to identify and safeguard biological corridors (hedges, forests, etc.) linking protected areas, so as to theoretically ensure the survival and adaptation of species to changes, owing to the exchange of individuals and therefore genes.

Greenways and Heritage Corridors: Land Use Planning and Landscape Ecology

Greenways are linear-shaped protected areas that are initially situated in the heart of or in proximity to urban areas. Greenways appeared in the United States in the 1970s, and increased in number from the end of the 1980s. According to Fabos (2004), the origin of greenways dates from the end of the 19th century, during which town planners imagined natural spaces within metropolitan open space systems. Subsequently, during the 1930s, the idea was to contain urban expansion by developing green lines inside or outside cities, or greenbelts, by relying on the local topography (mountains, rivers, etc.) to draw connection lines between these natural spaces. It appears that the concept of the greenway has been progressively used to specifically characterise spaces for the protection and tourist development of rivers and riverbanks. The term 'greenway' was used explicitly for the first time in 1987 by the President's Commission on Americans Outdoors[4], which laid down the framework for a greenway development programme by drawing a parallel with the American road (or rail) network. The objective was to create "a living network of greenways", a "giant circulation system".

Fabos and Ahern (1995) propose a typology of corridors which stems from this American movement. The first category consists of greenways which have a degree of ecological importance, concentrated along rivers, coastal areas or mountain ranges. Their objective is, on the one hand, to maintain biodiversity and the migration corridors of wild species, and on the other hand to restrict human activities, by acting as a containment barrier against urban pressure. The second category corresponds to recreational corridors. The idea here is to link various natural sites endowed with potential or effective tourist appeal. These recreational corridors can be situated in rural or urban areas. Finally, the third category of corridors refers to heritage corridors, i.e. sites with a high heritage value. Here, the purpose of this category of corridor is to offer a classification of the landscape based on the history of the economic and social relations between its various points. This type of corridor, as with the other two, is linear; most of the time consisting of rivers and riverbanks, even old roads, canals or railway lines that were used for important economic activity. The most famous heritage corridor is that of the Illinois and Michigan Canal that joins Lake Michigan in Chicago to the Illinois River, and therefore creates a corridor all the way to the state of Mississippi.

This fairly broad conception of corridors through the establishment of greenways is not restricted to Northern America; corridors have also been created in Europe and certain developing countries, such as China, where a National Green Corridor Programme was implemented in 1997 to "green" all roads (Yu et al. 2006).

This use of the corridor concept goes beyond the purpose of conservation, with the exception of ecological greenways, which have an obvious relationship with the conservation corridors examined previously. Heritage and recreational corridors are situated in a heritage, recreational and non-ecological context, which differentiates them from conservation corridors. Lastly, let us note that while the meaning of 'greenways' and 'heritage corridors' appears throughout the works of American urban architects from the end of the 19th century (Frederik Law Olmsted, George Kessler and Charles Elliot among others), these terms tend to take on a stronger geographic and institutional dimension from the end of the 1970s (Fabos 2004): 'geographic' because they are extended to a region, a federal state or even a country, and 'institutional' because governmental commissions and public-private

---

4   The Commission on Americans Outdoors was created by former American President Ronald Reagan in 1985, and entrusted to U.S. Senator Lamar Alexander. In his 1987 report, Alexander recommended networking recreational activities to enable people (pedestrians, cyclists, etc.) to circulate free of hindrance. The Commission's report is considered by many analysts as a major political event in terms of greenway promotion in the United States.

partnerships (such as the Chrysler Canada Greenway) are multiplying with a view to promoting the 'corridor' concept. The idea often put forward is to differentiate them from the classic parks, as managed by states, so as to promote alternative forms combining public and private funds, public spaces and private properties, and so on and so forth (Zubie 1995).

Towards the Infiltration of the Term 'Corridor' in Economics

The term 'corridor' is also found in economics as in 'development and/or transport corridor'. The parallel between conservation and development corridors is pertinent. Indeed, a development corridor is a communication route between at least two urban areas, and can involve various modes of transport (i.e. land, rail or river transport) for the transit of goods, workers and, potentially, economic information. Even if there are no specific definitions validated by economists – the literature on the subject being far less than that on ecological corridors – the concept of the development corridor also corresponds to a concern with increasing or improving the connectivity of flows (Arnold et al. 2005).

There was a considerable promotion of development corridors in the 1990s. This period was marked by an acceleration in the process of economic globalisation. The idea was to build large spaces in an economic context where exchange flows and the structuring of the largest international groups, led to the twin movements of globalisation and regionalisation. Furthermore, the transversal structuring of development corridors and their importance in relation to nation states, gave priority to transnational infrastructures, private actors and their affiliation to regional free trade setups.

For this reason, development corridors exist irrespective of the development status of the country or region concerned. Corridors are found in Europe (as with the European Backbone), as well as in North America (as with the North-Pacific Portland-Seattle-Vancouver corridor and the Californian San Diego-Los Angeles-San Francisco corridor) (Rimmer 1995).

Since the issue of development corridors is as diverse as the economic flows in question, a stricter definition of 'corridor' as exchange network structure is necessary. The corridor should be envisaged as the embodiment "of the passing of a firm's logic to the economy as a whole. In a given economy, all flows can be represented as deploying inside a spatial network comprising nodes, i.e. towns and metropolitan areas, and links corresponding to the different modes of transport and communications" (Rimmer 1995: 13). These development corridors are founded with the goal to reduce costs at city level.

Most of the corridors established in poor countries, such as those found in Africa, address a need to secure transport routes. They must be considered as a simpler form of the development corridor. They focus on the flow of goods between two or more points, often between a harbour and an urban area with no access to the sea. In fact, these corridors are often called 'transport' or 'transit corridors', the idea being that economic development in these countries cannot take place without an increased mobility in production factors.

The concept of 'corridor' in Economics can therefore take on various forms, from a simple vision, such as the transport corridor, where the emphasis is placed on the connectivity of towns (playing the role of a conduit for goods) with a strong territorial dimension, to the development corridor focused on the more or less complex networking of information flows. In the last case, territorial identity or geographic coherence is not essential, thus giving the impression of dealing with a 'paper' corridor existing only on maps, without any physical reality. In defining this type of corridor, Rimmer (1995) speaks of an "infrastructural scene".

Finally, one notes that development corridors, like conservation corridors, are rarely defined in an integrated manner. They do not take into account all the characteristics (e.g. cultural identities) or the scales required for land use planning.

Corridors: a Ragbag Concept

As we have just shown, corridors have many – and even sometimes diverging – definitions and functions. The absence of a clear and coherent terminology results in the actual objectives of corridors becoming confused (Simberloff et al. 1992; Bennett 1999). Concerning more specifically the conservation corridor, of which we have shown the significance in the field of conservation, we will see how the different definitions, concepts and expectations, as well as the lack of scientific conclusions, make conservation corridors barely workable in the context of biodiversity conservation.

Conduits or Habitats?

The movement of plants and/or animals (Hess and Fischer 2001) through a corridor is central to the majority of definitions: it is the function of conduit. Noss (1993) establishes that the two major functions of a corridor are to supply a habitat, in the sense of residence, and also to ensure a conduit for the purpose of movement. Rosenberg et al. (1995) separate clearly these functions of habitat and conduit. A corridor that enables travel between two patches, although it might not necessarily enable reproduction, represents the function of a conduit. When a corridor supplies the resources required for survival, reproduction and travel, it then plays the role of a habitat. As such there are ambiguities regarding the roles of conduit *versus* habitat when defining the function of a corridor. Indeed, some show that if a corridor constitutes a prime habitat for a species, this also facilitates the dispersion of that species (Bennett et al. 1994), and therefore its long term survival. Others focus on the conduit function and exclude from this concept spaces that constitute habitats but do not serve as conduits (Beier and Noss 1998). Nevertheless, in the 1990s a consensus existed among certain authors that the function of a corridor can range from a simple passage, to the role of both habitat and conduit (Hobbs 1992; Merriam 1991).

A Matter of Scale

Corridors also vary according to the time scales involved (Harris and Scheck 1991). 'Species' supposedly use corridors as conduits to move from one site to another, intermittently and over short periods of time, for very specific activities during their life (Beier and Loe 1992). This type of movement includes seasonal migrations, the daily search for food and journeys made for mating purposes (Noss 1991; Bennett et al. 1994). When a corridor is wide and long compared to the distances travelled by an animal, that species will use it over several generations. Beier and Loe (1992) call it a corridor dweller, and note that a corridor can be a habitat if it can support the reproduction of a species over several generations. Harris and Scheck (1991) link corridor width to usage type and duration. Individuals moving through narrow corridors do so in hourly or monthly time scales. Larger corridors support the movements of entire species over an annual cycle, and species assemblages can move through even larger corridors over decades or centuries. Narrower corridors can provide a habitat function because movements take place over several years. Movements within very large corridors concern whole communities and processes at the ecosystem level, enabling plant and animal species to travel between reserves over several generations. These have been called 'landscape linkages' and their purpose is to ensure regional connectivity (Noss 1991; Harris and Scheck 1991). Bennett (1999) prefers the term 'link' to that of 'corridor' to emphasise the conduit and landscape connectivity functions.

Little data is available to establish a link between these theories, the anticipated functions of conservation corridors and the creation of protected areas. The problem remains that the size of a corridor is closely dependant on the species under consideration and on the size of its territory. This is why conservationists work on the principle that conserving the largest territory belonging to a species enables the conservation of other species.

Conservation Corridors: from Theory to Practice

Obtaining operational data that can be of use to managers to delimit and manage conservation corridors is difficult, due to the diversity of corridor functions that varies by scale. Much confusion results partly from the double usage of the term 'corridor' on a structural and a functional level (Rosenberg et al. 1995). On the one hand, the connectivity provided by the corridor may be structural, a landscape linkage (Foreman 1995), and on the other it can be functional, contributing to the maintenance of meta-populations (Levins 1970; Hanski and Gilpin 1997; McCullough 1996). Baudry and Merriam (1988) distinguish structural from functional connectivity in that the linear elements of a landscape, which provide structural connectivity, do not automatically provide functional connectivity.

These definitions are particularly important when managers need to act. Indeed, when does one determine whether a process will have an impact on the functional connectivity of a corridor? The answer depends on its expected functions, the species involved, the time scales and the spaces considered.

These theoretical considerations conceal an even more complex reality as far as the efficiency of corridors for conservation is concerned. Corridors have positive effects on more than just species. They can conduct, slow down or even stop flows (Burel and Baudry 1999) and corridors, hedges and forests often come into conflict with the communication corridors established by humans (roads, paths, highways, sailing routes, etc.). Depending on the scale involved, these different corridors interact with a given species to constitute either a major route, or an insurmountable obstacle (See Figure 3.2).

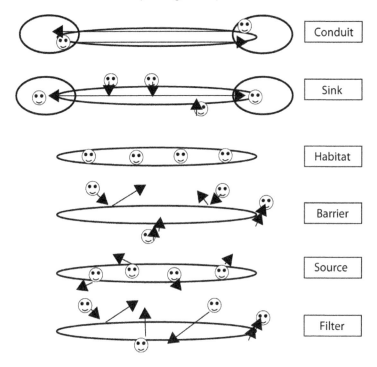

Source: Burel and Baudry (1999).

We can distinguish a bridge role (forests) for the passage of animals between two forest patches for example; a sink role (landscape element with a negative growth rate, therefore one which absorbs individuals) or a source role (with a positive growth rate which issues forth individuals); an ecological habitat role for species (a stream for a fish species); a barrier role (a river for terrestrial animals); a filter role, allowing the passage of certain species but not others.

Figure 3.2   The different roles of corridors depending on species and scales

A corridor which can be beneficial to the conservation of a species can also be detrimental to another. At this stage, we can already begin to comprehend the degree of complexity, and the potential conflicts, between what can be advantageous for one species and not for another; particularly when humans, as a species that also moves and builds communication routes for their own development, are a part of the system.

Lack of Data: A Source of Scientific Controversy

In practice, a significant amount of literature concerns the positive effects of corridors on animal flows, but far more rarely on the efficient flow of genes (e.g. the genetic homogeneity of a species along a corridor) which would enable a species to adapt over the long term. The many controversies shed light on the pernicious effects of corridors on species, populations and ecosystems. Many authors have in fact exchanged views in specialised journals regarding what Simberloff and Cox (1987), called the "consequences and costs of conservation corridors". These authors decided to raise the issues related to the lack of knowledge about the many effects of corridors, such as their importance in the transmission of pests, predators, diseases and bio-invasions, amongst others (Thomas et al. 2006). Notably, they questioned the balance between the ecological benefits and the (often considerable) economic costs related to the maintenance or implementation of corridors in order to save species inside and outside protected areas. One of their main arguments is that, in 1987, very little empirical data was actually available.

10 years later, Beier and Noss (1998) published a bibliographical review entitled *do habitat corridors provide connectivity?* While they were less clear-cut in their conclusions, they recognised that "generalisations about the biological value of corridors will remain elusive", particularly because of the fact that models depend on one species alone. As such, there is no clear answer regarding whether corridors maintain functional connectivity. However, they do note that, in 12 research articles, empirical works testify to the usefulness of corridors as conservation tools. Unlike those who are sceptical about corridors, Beier and Noss conclude that in the absence of valid data, and despite the high cost of these conservation actions, it is advisable to consider that a connected landscape is more desirable than a fragmented one. Therefore the precaution principle prevails in most conservation discourses and actions. As such, Beier and Noss (1998: 1250) address those who would contribute to the non-protection of these ecosystems by arguing that they "should bear the burden of proving that corridor destruction will not harm target populations".

Illustrating Corridor Challenges: the Case of Madagascar

Madagascar is the perfect example for understanding and analysing the process of implementing conservation corridors in a developing country. The dynamic set in motion by the Durban Congress played a part in the development of the Malagasy environmental policy. Indeed, it was on that occasion that the Malagasy President, Mr Marc Ravalomanana, declared that the country was to place 10% of its territory under protection, in order to meet international objectives. To this end, he proposed – in what is called in Madagascar the "Durban Vision"[5] – to triple the surface area of protected areas in the country within five years (Méral et al., this publication). Confronted with this particularly short deadline, urgency became the key word of all post-Durban conservation measures, while conservation corridors became the preferred tool for the creation of protected areas (Carrière-Buschsenchutz 2006).

Malagasy Corridors

The corridor concept came up in environmental policy debates in Madagascar during the *scientific workshop on the definition of the conservation priorities of biological diversity* in 1995. This concept, in a break from the model of protected areas classically applied in this country, is however perfectly adapted to the linear shape of the forest relicts (See Plate 11). Mixing the physical shape of the corridor with the ecological function of a conduit is an ideal reflection of the geographical reality of these forests. Within this framework, it was established that the forest 'corridors' would contribute towards establishing connectivity between protected areas, thereby playing a vital role in the maintenance of long term biodiversity (Carrière-Buschsenchutz 2006). These corridors are justified mainly by the connectivity they would ensure between protected areas, and also because the majority of the forests to be protected are situated within these forest strips. A major portion of the wooded area of Madagascar (around 50%, including currently protected areas) is affected, whether directly or remotely, by this corridor-centred approach. The evolution of conservation corridors (in red on plate, Plate 11), in relation to the remaining forest territories, testifies to their significance for conservationists. On such a scale, could all these corridors, were they contiguous, form regional landscape linkages that would be useful to the evolution of a species

---

5   "Durban vision" is a technical support group created by the Environment, Water and Forestry Directorate to implement the President's will via the System of Protected Areas in Madagascar (SAPM – système d'aires protégées malgache). Headed by the Secretary-General of the Department of Environmental and Water Affairs and Forestry, this group is made up of around 100 members representing more than 40 national and international organisations. The group on "Prioritisation" is responsible for proposing priority zones for the conservation of biodiversity, while the group on "Management and Legal Categorisation" is responsible for defining management objectives according to the potential categories of conservation areas.

over many generations? From a local corridor linking two protected areas, one moves here to a national system of meta-corridors, which logically has different expectations and objectives.

From the end of 2005, no less than one million additional ha were placed under protection – most by temporary decree – with 80% of these concerning forest corridors: the corridor of Anjozorobe-Angavo (52,000 ha), Ankeniheny-Zahamena (between 425,000 ha and 510,000 ha according to the sources) and the Makira Forest (around 350,000 ha). In the future, the surface area of the corridors of Eastern Madagascar ought to increase, since the corridors of Marojejy-Anjanaharibe-Sud (400,000 ha), Ranomafana-Andringitra-Midongy (240,000 ha), Tsitongambarika (147,000 ha), Marovoalavo (202,000 ha) and probably Fandriana-Marolambo (unknown surface area) should be added to those already established.

A Front for Conservation: From Political Choice to Theoretical Justification

The definitions, roles and expectations of corridors vary according to the actors and disciplines involved to form a fairly large overall concept. The surveys conducted in Madagascar lead to a similar conclusion. Depending on the interlocutors, the corridor – implicitly perceived in Madagascar as a forest corridor – is defined as a sort of "forest track", an "intermediary area" resulting from the wide expansion of a high priority ecosystem, a "biological bridge", a link between two protected areas, what remains of the forests, and even a "gene bank". Its function is also the subject of various interpretations, among which are the strategic role for the migration of species, the economic role of water tower for rice fields, a guaranty for genetic mixing, a natural protection for the species, a transition zone between two protected areas, a zone of sustainable management activity, and a forest full of natural resources, to name but a few. Certain conservation NGOs even integrate into their definition the idea that these corridors facilitate the creation of new protected areas, thereby ensuring the continuation of their activities.

Even if scientific results are lacking, the promotion of corridors in Madagascar is driven by good sense. All the scientists refer to the presumed role of corridors in the country by using the conditional tense (Carrière-Buschsenchutz 2006). All of them relate the controversies developed at the international level, as explained above. In Madagascar, the precaution principle largely justifies the conservation of these corridors, yet, these forest corridors are very rich in endemic species and this alone would be enough to justify their conservation. We can see here that while these forest strips could just as well be providing the functional role of a corridor, they represent excellent opportunities for conservation to successfully protect 10% of the land.

From being indispensable to the flow of genes, corridors have become indispensable to conservation policy in order to meet the challenge of the Durban vision. They went from species-rich ecological habitats to conduits for animals, which doubly justifies why they should be protected. The definition and delimitation of corridors are becoming redundant since, irrespective of what happens, the remainder of the Malagasy forests will have to be conserved. The function of corridor brings in an additional argument to justify conservation interventions, and especially to seek funds for their implementation. The proof being that the development plans of future conservation sites are not overly focused on the territories of a few key species using these corridors, but indeed on ecological forest or reef habitats (Chaboud et al., this publication), with all that they encompass.

Still, many scientists draw attention to the fact that each situation must be studied within the context of its specificity (Primack and Ratsirarson 2005). Some researchers have shown that species can react differently to the fragmentation of large forest blocs (Langrand and Wilmé 1997; Goodman and Rakotondravony 2000). Moreover, no development plan can provide an exhaustive and accurate report on the positive and negative effects (e.g. bio-invasion) expected from each one of these corridors. Recent studies have shown that the positive or negative role of corridors could be linked to the context and particularly to the frequency of disturbances. Indeed, when these are frequent, corridors can contribute to reducing the fixation of alleles beneficial to a species, whereas when they are rare, they increase it (Orrock 2005). Considering the extent of the disturbances on the Malagasy ecosystems (Goodman and Razafindratsita 2001; Lowry et al. 1997; Carrière and Ratsimisetra 2007) and the omnipresence of human activities in the remaining forests, we can ask whether it would not be relevant to test these hypotheses in the Malagasy context, before promoting the indiscriminate creation of protected areas covering all corridors.

All these studies only seem to justify further these conservation interventions when, for instance, there is a crucial need to integrate them into the other socioeconomic data, with a view to planning and conserving the land in a sustainable way or even develop it at the same time. The shortage of data should be an incentive to collect more, of better quality and on more relevant issues, rather than serve to make the argument against corridors (Carrière-Buchsenschutz 2006).

The Difficulty of Changing from Rhetoric to Practice: the Economic Argument

The economic justification for extending conservation corridors in Madagascar also reflects a gap between the political objectives and the efficient management of these corridors. Two arguments can be put forward.

Finance for the institutions responsible for the administration of these corridors is not guaranteed. The Malagasy Foundation for Biodiversity could have fulfilled this role, but it seems that the finance supplied by the trust fund will only just cover the recurrent costs of Madagascar National Parks, which manages 1.7 million ha of protected areas (Méral et al., this publication). The question of the financing of these corridors, which is reckoned to be $7 million for the first year, and $2 million of recurrent costs per year, is largely underestimated in the current debates and negotiations. One of the reasons for this is the possibility for the NGOs and the Malagasy state to resort to direct foreign financing (e.g. conservation contracts and private contributions, among others).

Furthermore, there remains the risk that the process for monitoring and managing the funds, assuming these cover the operating costs of the protected area, might keep the resident populations even further removed from the sources of financing. Indeed, the sums mentioned above concern only the operating costs of the institutions responsible for managing the protected areas, and not the opportunity costs endured by the local populations as a result of the restrictions thus created. What will be the compensation rules and measures for the locals? What shall constitute the ground rules for the maintenance of the forest cover which is required to obtain financing? All these questions, already the subject of debate among institutions responsible for securing permanent funding, are not tackled in the post-Durban deliberations. As an example, the surface of the core area of the Ankeniheny-Zahamena corridor (See Plate 11) is estimated to be 180,000 ha by the decree which established it. Confronted with such considerable surface areas, greater attention should be given to these crucial issues. Whereas evaluations performed on the existing network indicate a deficit in the compliance of the resident populations who only partially understand (and sometimes not at all) the advantages of conservation, the creation of vast protected areas – of the corridor type – appears somewhat irrelevant from the perspective of the challenges of sustainable development.

Conclusion

The corridor creation policies implemented in Madagascar are a good illustration of the persistent vagueness surrounding scientific knowledge and an actual corridor concept that could, under better circumstances, constitute an innovative rupture in the classic models of conservation.

Finally, although they may appear as novelties, corridors are part of a conservationist strategy based on a top-down approach, within which sites are identified only according to ecological criteria, with the sole intention of increasing the extent of protected areas. A bottom-up approach would integrate the human factor, with its social and cultural values, into the implementation of new corridors, thereby improving their management methods and efficiency in terms of both conservation and sustainable development (cf. a formulation of the reticular model in Albert et al., this publication).

By trying to apply the corridor concept in countries with very different levels of development, there is a considerable risk that greatly contrasting results, or even undesirable and unexpected side-effects, will be obtained. Implementing and expanding corridors within the framework of the Pan-European Ecological Network (Bonnin, this publication) for example, does not address the same situation and constraints as in a country of great poverty, such as Madagascar. The objectives emanating from the Durban Congress, while they appear pertinent when viewed from the perspective of area protection networking and conservation actors, can remain difficult to implement in developing countries.

References

Arnold J., Ollivier G., Arvis J. F., 2005 – *Best practices in corridor management.* Washington, World Bank.

Baudry J., Merriam H. G., 1988 – Connectivity and connectedness: functional versus structural patterns in landscapes. *Münstersche Geographische Arbeiten*, 29: 23–28.

Beier P., Loe S., 1992 – A checklist for evaluating impacts to wildlife movement corridors. *Wildlife Soc. Bull.*, 20: 434–440.

Beier P., Noss R., 1998 – Do habitat corridors provide connectivity? *Conservation Biology*, 12: 1241–1252.

Bennett A. F., 1999 – *Linkages in the landscape. The role of corridors and connectivity in wildlife conservation.* Gland, Cambridge, IUCN, p. 254.

Bennett A. F., Henein K., Merriam G., 1994 – Corridor use and the elements of corridor quality: chipmunks and fencerows in a farmland mosaic. *Biol. Conservation*, 65: 155–165.

Blondel J., 1995 – *Biogéographie. Approche écologique et évolutive*, p. 297. Paris/ Milan/Barcelone, Éditions Masson, Coll. Écologie n° 27.

Blondel J., Perret P., Maister M., Dias P., 1992 – Do harlequin Mediterranean environments function as source-sink for Blue Tits (*Parus caeruleus* L.)? *Landscape Ecology*, 6: 213–219.

Boorman S. A., Levitt P. R., 1973 – Group selection on the boundary of a stable population. *Theoretical Population Biology*, 4: 85–128.

Burel F., Baudry J., 1999 – *Écologie du paysage. Concepts, méthodes et applications*. London/New York/Paris, Éditions Technique et Documentation, p. 359.

Carrière S., Ratsimisetra L., 2007 – "Le couloir forestier de Fianarantsoa : forêt primaire ou forêt des hommes". *In* Serpantié G., Carrière S., Rasolofoharinoro M. (eds.), *Transitions agraires, dynamiques écologiques et conservation. Le corridor de Fianarantsoa*. Workshop Gerem CNRE-IRD, Antananarivo : 39–46.

Carrière-Buchsenschutz S., 2006 – L'urgence d'une confirmation par la science du rôle écologique du corridor forestier de Fianarantsoa. *Études Rurales*, Special issue, 178: 181–196.

Fabos J. G., 2004 – Greenway planning in the United States: its origins and recent case studies. *Landscape and Urban Planning*, 68: 321–342.

Fabos J. G., Ahern J. (eds.), 1995 – *Greenways. The beginning of an international movement*. Amsterdam, Elsevier, p. 498.

Fahrig L., Merriam H. G., 1985 – Habitat patch connectivity and population survival. *Ecology*, 66: 1762–1768.

Forman R. T. T., 1995 – *Land mosaic. The ecology of landscapes and regions.* Cambridge, Cambridge University Press, p. 656.

Forman R. T. T., Godron M., 1981 – Patches and structural components for a landscape ecology. *BioScience*, 31: 733–740.

Forman R. T. T., Godron M., 1986 – *Landscape ecology.* New York, John Wiley and sons, p. 640.

Goodman S. M., Rakotondravony D., 2000 – The effects of forest fragmentation and isolation on insectivorous small mammals (*Lipotyphla*) on Central High Plateau of Madagascar. *Journal Zoological*, 250: 193–200.

Goodman S. M., Razafindratsita V. R., 2001 – *Inventaire biologique du parc national de Ranomafana et du couloir forestier qui le relie au Parc national d'Andringitra.* Antananarivo, CIDST, Recherches pour le Développement n° 17, p.243.

Hanksi I., Gilpin M. E. (eds.), 1997 – *Metapopulation biology, ecology, genetics and evolution.* San Diego, Academic Press, p. 512.

Harris L. D., Scheck J., 1991 – "From implications to applications: the dispersal corridor principle applied to the conservation of biological diversity". *In* Saunders D. A., Hobbs J. (eds.), *Nature conservation 2. The role of corridors.* Chipping Norton, Surrey Beatty and Sons: 189–220.

Hess G. R., Fischer R. A., 2001 – Communicating clearly about conservation corridors. *Landscape and Urban Planning*, 55: 195–208.

Hilty J. A., Lidicker Jr. W. Z., Merenlender A. M., 2006 – *Corridor ecology. The science and practice of linking landscape for biodiversity conservation.* Washington, Island Press, p. 344.

Hobbs R. J., 1992 – The role of corridors in conservation: solution at bandwagon. *Trends Ecol. Evolution*, 7: 389–392.

Langrand O., Wilmé L., 1997 – "Effects of forest fragmentation on extinction patterns of the endemic avifauna on the central high plateau of Madagascar". *In* Goodman S. M., Patterson B. D. (eds.), *Natural change and human impact in Madagascar.* Washington/London, Smithsonian Institution Press: 280–305.

Levins R., 1969 – Some demographic and genetic consequences of environmental heterogeneity for biological control. *Bull. Entomol. Soc. Am.*, 15: 237–240.

Levins R., 1970 – "Extinction". *In* Grestenhaber M. (ed.), *Some mathematical questions in biology. Lectures on mathematics in the life sciences.* Providence, American Mathematical Society: 77–107.

Lowry II P. P., Schatz G. E., Phillipson P. B., 1997 – "The classification of natural and anthropogenic vegetation in Madagascar". *In* Goodman S. M., Patterson B. D. (eds.), *Natural change and human impact in Madagascar.* Washington/London, Smithsonian Institution Press: 93–123.

McArthur R. H., Wilson E. O., 1967 – *The theory of island biogeography.* Princeton, Princeton University press, p. 230.

McCullough D. R. (ed.), 1996 – *Metapopulations and wildlife conservation.* Washington, Island Press, p. 439

Merriam G., 1991 – "Corridors and connectivity: animal populations in heterogeneous environments". *In* Saunders D. A., Hobbs J. (eds.), *Nature conservation 2. The role of corridors*. Chipping Norton, Surrey Beatty and Sons: 133–142.

Noss R. F., 1991 – "Landscape connectivity: different functions at different scales". *In* Hudson W. E. (ed.), *Landscape linkages and biodiversity*. Washington, Island Press: 27–39.

Noss R. F., 1993 – "Wildlife corridors". *In* Smith D. E., Hellmund P. C. (eds.), *Ecology of greenways. Design and function of linear conservation areas*. p. 294. Minneapolis, University of Minnesota Press: 43–68.

Orrock J. L., 2005 – Conservation corridors affect the fixation of novel alleles. *Conservation Genetics*, 6: 623–630.

Primack R. B., Ratsirarson J., 2005 – *Principe de base de la conservation de la biodiversité*. Antananarivo, MacArthur Foundation/ESSA/CITE.

Pulliam H. R., 1988 – Sources, sinks, and population regulation. *American Naturalist*, 132: 652–661.

Rimmer P. J., 1995 – La nouvelle "scène infrastructurelle" de l'Asie Pacifique; son émergence depuis le début des années soixante-dix. *Revue 2001 Plus,* 35: 7–43.

Rosenberg D. K., Noon B. R., Meslow E. C., 1995 – "Towards a definition of biological corridor". *In* Boissonette J. A., Krausman P. R. (eds.), *Integrating people and wildlife for a sustainable future*. Bethesda, The Wildlife Society: 436–439.

Simberloff D., Cox J., 1987 – Consequences and costs of conservation corridors. *Conservation Biology*, 1 (1): 63–71.

Simberloff D., Farr J. A., Cox J., Mehlman D. W., 1992 – Movement corridors: conservation bargains or poor investments. *Conservation Biology*, 6: 493–504.

Thomas J. R., Middleton B., Gibson D. J., 2006 – A landscape perspective of the stream corridor invasion and habitat characteristics of an exotic (*Dioscorea oppositifolia*) in a pristine watershed in Illinois. *Biological Invasions*, 8: 1103–1113.

Yu K., Li D., Li N., 2006 – The evolution of Greenways in China. *Landscape and Urban Planning*, 76: 223–239.

Zube E. H., 1995 – Greenways and the US national park system. *Landscape and Urban Planning,* 33: 17–25.

Chapter 4

Protected Areas and Ecological Networks: Global Environmental Management or Management of the Conservation Institutions?

Marie Bonnin

Using the ecological network concept, it might possible to achieve sustainable development via territorial zoning (Bonnin 2008; Cedre 2002). Of course, merely identifying zones is insufficient and requires further investigation from a legal point of view in order to understand the regulatory implications of zoning. In other words, it is important to question the various standards applicable to different zones. To what extent is a strict form of conservation still useful? Or in which situations should one reduce constraints and use more incentive instruments, i.e. ones that might lead to the integration of nature conservation into sectoral policies, such as transport or agriculture? Thus, the aim of this chapter is to evaluate the impacts of the emergence of the ecological network concept has had on protected areas. One such impact is that protected areas no longer play the exclusive role of nature and biodiversity conservation.

From a conceptual point of view, ecological networks are often apprehended through the use of a system that represents the three types of areas most commonly utilised in establishing ecological networks. These are the 'core zones', 'buffer zones' and 'biological corridors'[1] (Jongman and Pungetti 2004; Sepp and Kaasik 2002; Carrière et al., this publication). We need therefore to place current and future protected areas into this system.

Although we have intentionally chosen to adopt an internationalist approach here, it is important to highlight the importance of national protected areas before limiting this analysis to natural areas protected by an international classification. Most countries have developed their own protected area classification system, which can range from strict protection systems, as in the case of integral

---

1   For example, Estonia uses the terms 'core zones', 'buffer zones' and 'ecological corridors', Lithuania the terms 'geosystems', 'buffer territories' and 'ecological corridors', Poland uses the terms 'core zones' and 'eco-corridors'. The Slovak and Czech networks are made up of 'bio-centres', 'bio-corridors' and 'interactive elements'. The Netherlands use the terms 'core zones', 'reconstitution area' and 'ecological corridors'.

Table 4.1   Various international conservation area networks

Denomination	Date of creation	Responsible Organisations	Objectives	B or NB*	Scope	State obligations	Terms and conditions of site designation
European Diploma	6 March 1965	Council of Europe	To protect exceptional and particularly well protected sites	NB	Europe	To maintain the protection level	On the proposal of governments after permission and consent from an expert committee
Ramsar Sites	2 February 1971	Secretariat of the Convention, provided by the IUCN	To conserve wetlands	B	World	To create reserves to conserve wetlands	On the proposal of governments
World Heritage Sites	23 November 1972	UNESCO	To conserve natural heritage of exceptional universal value	B	World	To actively ensure the protection, conservation and development of the heritage concerned	On the proposal of governments after approval by an intergovernmental committee
Biosphere Reserves	1976	UNESCO	To conserve natural habitats; to encourage research	NB	World	To elaborate on appropriate zoning and management plans	On the proposal of governments and European Council experts
Biogenetic Reserves	15 March 1976	Council of Europe	To protect samples representative of the natural heritage; to promote research; to heighten public awareness	NB	Europe	To ensure that the protection status is compatible with the objectives of the area	On the proposal of governments

Mediterranean Specially Protected Areas	1982	Regional Activity Centre for Mediterranean Specially Protected Areas	To conserve natural areas	B	Mediterranean Sea Region	To adopt common criteria in creating and managing the areas	On the proposal of governments
Natura 2000	21 May 1992	European Union	To conserve natural habitats	B	European Union	To protect the habitats listed in the Annexes	On the proposal of governments
System of Coastal and Marine Baltic Sea Protected Areas	1994	Helsinki Commission	To conserve natural areas	NB	Baltic Sea Region	Not yet defined	On the proposal of governments
Emerald Network	1996	Council of Europe	To conserve natural habitats	NB	Bern Convention (Europe)	Not yet defined	On the proposal of governments

Remark: Binding (B) or Non-Binding (NB). 'Binding' refers to the binding force of the legal instrument behind conservation area networks.

nature reserves or certain national parks, to less restrictive protection systems, such as landscape parks, via the targeted and systematic protection of certain habitats or ecosystems (Rodary and Milian, this publication). The introduction of international and European regulations on nature and biodiversity protection, encouraged governments to launch co-ordinated actions to identify and resolve, at the supranational level, the major problems posed by conservation.

As a result, various declarations and legislation[2] implement ecological networks, and it is important to question the place and role of protected areas in these integrated conservation systems (See Table 4.1).

Some international organisations liken protected areas to the core zones of ecological networks. In this regard, the core zones of the Alpine ecological network established by the Alpine Convention Secretariat, are made up of large protected areas (Alpine Network 2004). The Council of Europe, in supporting the creation of the Pan-European ecological network, considers that the core zones of this network are made up of exceptionally valuable natural areas, whether they are to be protected, or are already protected. Other ecological networks being implemented on different scales may have adopted other definitions (Bennett and Wit 2001), and the place of protected areas in the elaboration of territorial zonings remains antagonistic in the absence of consensus on this matter (Carrière et al., this publication). We nevertheless show here that most political decisions and legal texts on the implementation of ecological networks, agree on the role of protected areas as core zones, although in some cases they can also act as corridors.

Protected Areas as Core Zones

Areas protected by international regulations have been multiplying since the 1970s (Rodary and Milian, this publication). Implemented within the framework of a binding international agreement, for example the World Heritage Convention (Paris, 1972[3]), or a non-binding one, such as the resolution of the Committee of Ministers establishing the European Diploma, these protected areas are included in a list, without prejudice to state sovereignty (Kiss and Beurier 2004). States, by adhering to the international convention, accept to take on responsibilities vis-à-

---

2   There is a multiplication of political declarations anticipating the creation of ecological networks: the Pan-European Strategy for Biological and Landscape Diversity in Europe (Sofia, 1995), the Strategy for Biodiversity and its Plan of Action for Central Asia (2006). Several binding texts in international law exist now on the subject: the Carpathian Convention and its article 4 (Kiev, 2003) in particular, and the Convention for the Conservation of the Biodiversity and the Protection of Priority Wilderness Areas in Central America (Managua, 1992).

3   The exact references of the international conventions quoted are grouped together at the end of this text in Table 4.3. The chapter only mentions the place and date of signature.

vis the protection of these natural areas. The Council of Europe, a pioneer in the field, has since 1965 maintained a list of protected sites based on the awarding of the European Diploma for Protected Areas. Other organisations rapidly followed suit by also opting for the listing of natural areas (Table 4.1), such as UNESCO with its 'World Heritage Sites' and 'Biosphere Reserves'. A site being entered on a list of sites means there will be an exchange of information and best practices between the area administrators within the international organisation in charge, via periodic meetings in particular. This networking facilitates the transfer of knowledge, which can be stimulating for the various actors. However, the use of the term 'networks of protected areas' or 'network of sites' is only justified by these exchanges between area administrators within the international organisation. With the exception of a few bird species, for which an ecological network could consist of unconnected islands of nature, ecological networks imply a territorial continuity that bears little relation to the networks of protected sites at the international level, such as the network of UNESCO's World Heritage Sites.

We show here that while these different lists of protected areas all intend essentially to conserve biodiversity, each targets specific objectives. The evolution of protection methods inside and outside these sites leads to the integration of secondary objectives that also aim to protect 'ordinary nature'.

Common but Differentiated Objectives

Certain conservation area networks aim at safeguarding and protecting exceptional sites. This is particularly the case for the World Heritage Site network, and for the sites of the European Diploma of Protected Areas, which intends to reward the exemplary management of natural or semi-natural areas or landscapes. This represents an exceptional level of European interest in their biological, geological and landscape diversity. Other conservation area networks have more targeted objectives for the protection of natural areas with high ecological stakes. The Ramsar sites network for example aims at protecting a type of ecosystem which is particularly rich biologically, but is also globally endangered. Ramsar sites play a fundamental role in protecting the migratory routes of water birds, as well as in the proper management of the ecological processes and functions of wetlands. Thus, they are involved in the implementation of a coherent spatial structure that plays a specific role in flood prevention, and mitigating the impact of pollution. In this respect, the Ramsar sites fulfil the sustainable development objectives included in the implementation of an ecological network.

Whether it is binding or not, the legal basis of these conservation area networks affects the way protected areas will protect biological diversity, in the sense that this legal basis will not have the same status before the various jurisdictions of the contracting states (Romi 1990). Some conservation area networks are created through a binding international convention, as is the case with the Ramsar network. Others are founded on the basis of a simple resolution or recommendation that does not have a legal value in the strict sense of the word, as is the case for the

European Diploma. Furthermore, the binding aspect of these state obligations results not only from their binding force, but also from their objectives. Indeed, the value of state obligations also depends on the possibility a government has to remove a zone from the network. Yet, in several conservation area networks, governments can still do so without having to justify themselves. With regard to biosphere reserves, a state can remove an area from the network by simply notifying the secretariat. The Ramsar Convention, in Article 2, Paragraph 5, allows the removal of a site from the list of internationally important wetlands, provided there are urgent national interests, and prior information to the bureau in Gland (Switzerland). There is also an obligation to compensate for the loss of wetland resources (De Klemm 1998). The impact of these provisions is important, for including a site in a network equates to protecting it from major and mainly state-led infrastructural projects. Yet, if a state can remove a site from the network at any time, whether or not the obligations are binding, has no influence on the effectiveness of the protection. However, the diplomatic value of a listed site is such that governments almost never remove a site from the network.

The inclusion of protected areas as part of a network of international sites, beyond the legal protection it offers, also allows a supra-national value to be given to certain natural habitats, which in turn makes it possible to appreciate differently certain development projects, when evaluated in relation to a natural area of international importance. Protection then becomes more political than legal. In fact, the sanction mechanisms against a contracting state misusing a protected area go as far as removing a zone from the network. These mechanisms are not without impact; for example we can mention the Whale Sanctuary of El Vizcaino, in Mexico. In 1999, a project for the expansion of a salt production plant was launched in the San Ignacio Lagoon, in the Bay of Sebastian Vizcaino, the last intact lagoon where the Pacific gray whale comes to breed. This site is included in the World Heritage List. The World Heritage Committee warned the Mexican government of the threat a salt plant inside the sanctuary would pose to the marine and terrestrial ecology, the Pacific gray whales and the integrity of the site. In March 2000, the Mexican government decided not to give permission for the construction of the salt plant.

Internationally, protected areas appear particularly important for their ability to protect specific natural habitats. Their role as core zones of ecological networks is, however, also under transformation, and this can be noticed in the evolution of the terms and conditions of protection in certain conservation area networks.

Ecological Networks within Protected Areas

Certain large protected areas integrate the concept of the ecological network into their territory, and can therefore act as an 'experimental laboratory' by modifying the classic approach of conservation. Such is the case of the biosphere reserves that, since the adoption of the Statutory Framework (UNESCO 1996) and the Seville Strategy, have three clearly defined functions. These are described as being complementary and of equal importance: a function of conservation (protecting genetic resources, areas and ecosystems as well as landscapes), development (fostering sustainable economic and human development) and logistics (enabling and fostering research, permanent monitoring, educational and training activities).

To these ends, they are divided into three types of zones: a central area endowed with a legal status guaranteeing long term protection, and in which most human activity is forbidden; a clearly defined buffer zone, in which only those activities compatible with the conservation objective are authorised; and a transition area which does not normally have a protected status, and which allows and favours the sustainable use of resources (Cibien 2006). Today, this zoning, which was officialised in 1995 by the Statutory Framework and the Seville Strategy, is expressly used to set up biological corridors inside biosphere reserves. The recent zoning of the East Carpathian Biosphere Reserve is a case in point. It does not seem necessary to extend to all protected areas this tendency to set up ecological networks inside the territory of the protected natural area. The size of the protected areas, as implied by this process, can indeed limit the public's understanding and acceptance of protected areas without the creation of tangible benefits. Moreover, certain lists of protected areas at the international level have been established to strictly protect particular areas, and are not well adapted to this type of zoning which integrates economic activity. This is particularly the case for the World Heritage List, or the sites of the European Diploma. This situation does not help their complementarity with other lists of sites, or improve the comprehension of their processes by the public[4]. On the other hand, the biosphere reserve concept as it applies to national legislations and which favours conciliatory measures in addition to economic development (while maintaining areas of strict conservation), might represent a means to achieve the sustainable development of the territory, by enabling the conservation of ordinary nature on the outskirts of protected areas.

Linking Protected Areas together Implies their Conservation

Conserving nature through protected areas also has side effects for the management of ordinary nature outside the land of the protected area considered. Indeed, several international texts simply request that contracting states link protected areas to one another.

Made up of special protection areas under the Birds Directive, and special areas of conservation under the Habitats Directive, the Natura 2000[5] network did

---

4   Although the Val de Loire which was included in the World Heritage List on the 30th of November 2000 remains an exception, it still illustrates this phenomenon. The perimeter goes from Sully-sur-Loire (upstream) to Chalonnes-sur-Loire (downstream), i.e. an area 260 km long and a few kilometres wide, corresponding to the main riverbed, and including 159 *communes*.

5   In December 2006, the Natura 2000 network included 20,862 sites under the Habitats Directive, of which 1,250 were marine sites, and 4,617 sites under the Birds

not initially aim to implement an ecological network (in the sense described in the introduction). However, the final creation of the network implies a certain ecological coherence, as indicated by several Directive articles. The progress achieved in fulfilling the objectives of the Natura 2000 network now enables the European Commission to begin envisaging the next step, as indicated in its 2006 communication on biological diversity[6]. The first implementation phase was focused on the proposal and designation of sites hosting species and habitats of continental interest by the member states of the European Union. The next steps of the Natura 2000 network will aim to ensure the operational character of the network, particularly to ensure that species and habitats of European importance are maintained in a favourable state of conservation. Establishing the conservation measures necessary for all the designated sites, including the elaboration of management plans, the adoption of an appropriate national status, as well as administrative or contractual measures, now represents one of the priority tasks of member states.

In the context of climate change and land use transformations, the capacity of the network to meet its conservation objectives depends particularly on the maintenance or restoration of an appropriate matrix of landscapes, both in and between the sites, making it possible to maintain essential ecological processes and favouring biodiversity. With reference to Article 10 of the Habitats Directive, the European Commission and certain member states are currently endeavouring to define the conditions and resources necessary for ensuring, within the Natura 2000 network, the coherent management of landscape features (Box 4.1).

Other international texts on the protection of nature also highlight the necessity to link protected areas. The Convention on Biological Diversity (Rio, 1992) is certainly the best example in this regard. According to the terms of Article 8 (a), "Each Contracting Party shall […] establish a system of protected areas or areas where special measures need to be taken to conserve biological diversity". The Convention did not adopt the expression 'ecological connectivity' in the final text, but instead adopted 'system' which is typically associated with the analysis of the preparation works, and therefore suggests that the Parties should be involved in establishing connected protected areas[7].

Certain regional conventions specifically ask for the establishment of ecological corridors. The Central American Convention for the Protection of the Environment pointed out, as early as 1992, the importance of the Central American isthmus as a biological corridor. In 1994, the Alpine Convention also recognised the importance

---

Directive, of which 484 were marine sites.

    6   "Halting the loss of biodiversity by 2010 and beyond. Sustaining ecosystem services for human benefit", Commission communication, COM (2006) 216, May 2006.

    7   The guide of the Convention on Biological Diversity in fact, based on Article 8(a), advocates "Establishing a larger protected area estate, than would otherwise be the case, with emphasis on creating corridors and "stepping-stones" between protected areas so as to enable species to move with shifting climate".

Box 4.1   Direct legal references to ecological coherence in the Habitats Directive

---

Preamble

[I]n order to ensure the restoration or maintenance of natural habitats and species [...] at a favourable conservation status, it is necessary to [...] create a coherent European ecological network [...].

Article 1

(k) *[S]ite of Community importance*: a site which, [...] may also contribute significantly to the coherence of Natura 2000 referred to in Article 3, and/or contributes significantly to the maintenance of biological diversity within the biogeographic region or regions concerned.

Article 3

3. [...] Member states shall endeavour to improve the ecological coherence of Natura 2000 by maintaining, and where appropriate developing, features of the landscape which are of major importance for wild fauna and flora, as referred to in Article 10.

Article 4

4. [...] [T]he Member state concerned shall designate that site as a special area of conservation [...] establishing priorities [...] for the coherence of Natura 2000 [...].

Article 10

Member states shall endeavour, where they consider it necessary, in their land-use planning and development policies and, in particular, with a view to improving the ecological coherence of the Natura 2000 network, to encourage the management of features of the landscape which are of major importance for wild fauna and flora.

---

of establishing an ecological network in the Alps, via its protocol for nature and landscape conservation. And the recent Framework Convention on the Protection and Sustainable Development of the Carpathians, signed in Kiev in 2003, defining a legal framework for the sustainable protection of its ecosystems, invited its contracting Parties to take appropriate measures with a view to ensuring a high level of protection of natural and semi-natural habitats, as well as their continuity and connectivity (Fall and Egerer 2004). This convention refers explicitly to the necessity for the Parties to constitute an ecological network in the Carpathians, which implies the creation "of a network of protected areas associated with the conservation and sustainable management of the areas outside of protected areas". Nonetheless, seeking to ensure the continuity of protected areas does not imply their abusive expansion. In fact, the challenge of the connectivity concept is to associate other, more flexible, methods of protection with protected areas.

In addition to protecting defined natural habitats, protected areas can have an impact outside their own space, which contributes to the development of their role regarding biological diversity conservation. The fact that a structure exists can also serve as logistical support for initiatives on the outskirts of the central area. As such, although their role always appears to be necessary, they must be perceived differently according to the scale of reflection.

Protected Areas as Corridors

Certain protected areas may play the role of a 'corridor', i.e. making it possible to reconnect or connect several populations of species by protecting natural infrastructures. Already at the end of the 19th century, a notion inspired by natural linear features had been proposed by American planners. This resulted in the concept of greenways (Carrière et al., this publication). Greenways are networks of linear features, planned and arranged initially for recreational purposes (Walmsley 2006; Fabos and Ryan 2004), but have since accumulated multiple goals, as well as recreation, including ecology, aesthetics and cultural spaces (Ahern 1995). Greenways do not only have an ecological function, they are multifunctional ecological networks too. Most of them are situated around towns, and aim to make it easy for city dwellers 'to get some fresh air', as well as to maintain a natural landscape around large urban structures. On the European continent, the idea of the ecological network initially took another form, and was first implemented by town planning officials in Russia, Czechoslovakia and Lithuania. They integrated the corridor-tool into their town planning systems during the 1970s, in order to protect natural infrastructures (Jongman 1998). This approach relied on the concept of a polarised landscape, which implies the fragmentation of the landscape into areas for the conservation and restoration of nature, and areas for intensive land use (Frolova 2000; Kavaliauskas 1996). This tradition of natural infrastructure planning explains the prevalence of environmental law pertaining to ecological networks in these countries[8]. More recently, Western Europe took into account the importance of reducing natural habitat fragmentation (Burel 2003), and several countries now have legal texts implementing ecological networks[9].

These historical differences in the conception of ecological networks explain how, depending on its location, the term 'corridor' can take on different meanings. Naturally these terminological variations are a source of confusion (CBD 2005; Bennett and Wit 2001; Carrière et al., this publication). While Europe and international organisations normally use 'ecological network', South American or Asian programmes generally use 'corridor', which nonetheless corresponds to

---

8   For example, Estonia (the Sustainable Development Act of 1995), Lithuania (the Environmental Protection Act of 1992), the Czech Republic (the Nature Protection Act of 1992) and Slovakia (the Nature Protection Act of 1994).

9   For example, Germany (the Nature Conservation Federal Act of 2002), Belgium, Flemish Region (the Decree on Nature Conservation of 1997), France (the Orientation Act of 1999 on the Sustainable Development of the Territory), and Switzerland (Landscape Conception, 1997).

the same model of conservation[10]. The approach taken by the Conference of the Parties to the Convention on Biological Diversity will be adopted here: the idea of an ecological interconnection is regarded as a 'corridor'. More specifically, the term 'corridor' specifies one or more environments functionally linking different habitats vital to a species or a group of species.

Transboundary Protected Areas as Support for Interregional Corridors

Transboundary protected areas play a particular role in the regional dynamic of nature conservation (Brunner 2002). Making a transboundary protected area part of a network of international sites not only encourages institutional contact between the officials of areas situated on either side of the border, it also bases co-operation within a legal and political framework that can lead to more general initiatives.

Many international nature conservation conventions impose on their signatories a duty to co-ordinate their actions in the field of transboundary protected areas[11], and Party conferences now recommend common management measures[12]. The legal recognition of transboundary protected areas is a first step towards legally recognising that regional co-operation is necessary regarding nature conservation. The fact that several conservation area networks have allocated unique names to protected areas on either side of a border confirms this. This is particularly the case for the Bialowieza Forest, between Poland and Belarus, which was designated as such by the World Heritage Sites and the European Diploma.

One of the main objectives of the Seville Strategy for Biosphere Reserves (UNESCO 1996; Jardin 1996) is to promote the twinning of biosphere reserves, and encourage the creation of transboundary reserves. Internationally designated transboundary protected areas have been multiplying these last years (Table 4.2).

The establishment of joint management commissions which include representatives from each protected area also constitutes a step towards the establishment of co-operation mechanisms on a regional scale. It is interesting to note that, often, the establishment of such commissions coincides with the establishment of corridor protection dynamics on an interregional scale. This is particularly the case for La Amistad National Park between Costa Rica and Panama, for which a permanent bi-national commission has been established, chaired by

---

10   Such as the Vilcabamba-Amboro Conservation Corridor between Peru and Bolivia, and the Mesoamerican Biological Corridor in Central America.

11   This is the case, for example, of the Bern Convention on the Conservation of European Wildlife and Natural Habitats of 1979, the ASEAN Agreement on the Conservation of Nature and Natural Resources (Kuala-Lumpur, 1985) or the Agreement on the Conservation of African-Eurasian Migratory Waterbirds (The Hague, 1995).

12   The second strategic plan of the Ramsar Convention on wetland conservation (2003–2008) indicates that the Parties ought to co-operate internationally in their delivery of transboundary wetland conservation and their wise use.

Table 4.2   Transboundary protected areas recognised by international denominations

Date of creation	Name of protected area	Countries concerned	International denomination
1973	German-Luxembourg Nature Park	Germany, Luxembourg	European Diploma
1982	Belovezhskaya Pushcha / Białowieża Forest	Belarus, Poland	World Heritage Sites European Diploma
1982	Mount Nimba Strict Nature Reserve	Côte d'Ivoire, Guinea	World Heritage Sites
1989	Mosi-Oa-Tunya	Zambia, Zimbabwe	World Heritage Sites
1990	Talamanca Range-La Amistad Reserves	Costa Rica, Panama	World Heritage Sites
1992	Tatra	Poland, Slovakia	Biosphere Reserves
1992	Krkokonose/ Karkonosze	Czech Republic, Poland	Biosphere Reserves
1994	Kluane/Wrangell-St Elias/ Glacier Bay/ Tatshenshini Alsek	United States, Canada	World Heritage Sites
1995	Waterton Glacier International Peace Park	United States, Canada	World Heritage Sites
1998	Pfälzerwald-Voges du Nord	France, Germany	Biosphere Reserves
1998	Danube Delta	Romania, Ukraine	Biosphere Reserves
1998	East Carpathians	Poland, Slovakia, Ukraine	Biosphere Reserves
1999	Pyrénées – Mont Perdu	Spain, France	World Heritage Sites (mixed landscapes)
2000	Caves of Aggtelek Karst and Slovak Karst	Hungary, Slovakia	World Heritage Sites
2000	Courland Isthmus	Lithuania, Russia	World Heritage Sites
2002	W Regional Park	Benin, Burkina Faso, Niger	Biosphere Reserves
2003	Uvs Nuur Basin	Russia, Mongolia	World Heritage Sites
2005	Senegal River Delta	Mauritania, Senegal	Biosphere Reserves
2006	Kvarken Archipelago	Finland, Sweden	World Heritage Sites

the Ministers of Planning. This commission is responsible for the programme, the projects and the co-ordination of the general activities, as well as for their monitoring and evaluation. The first co-operation agreement was in fact signed in 1979, between Panama and Costa Rica. The creation of La Amistad National Park was ratified by an agreement signed in 1982 and, in 1992 the common advisory commission became a permanent commission, and acquired the power of decision making. This transboundary park is at the heart of an international initiative, the Mesoamerican Biological Corridor. Another example would be the transboundary nature protection area in the nature reserves of the Danube Delta, which was the subject of an agreement concerning the establishment of a greenway in the lower Danube (Bucharest, 2000).

Transboundary protected areas therefore have the potential to play a role in the establishment of inter-regional corridors, particularly via the legal basis they establish, and consequently can procure a basis for a regional conservation policy dynamic.

Archipelagos of Protected Areas as Corridors on a Larger Scale

Depending on their scale, a myriad of protected areas can also constitute a corridor. This would be the case, for example, of a protected area situated on farm land, intended for protecting an otter or some other protected animal. This area is perceived at the level of the regional development system in the same way as protected areas situated on neighbouring agricultural lands, as a biological corridor.

Similarly, the Alpine ecological network, made up of protected areas and implemented as part of the Alpine Convention (Salzburg, 1991), is acknowledged by various studies for its role of corridor. The establishment of spatial links between protected alpine areas is a central theme of this convention, and in particular its nature and landscape conservation protocol (Chambéry, 1994), which contains the Article 12, entitled *Ecological Network*. The Parties to this convention have highlighted that only large protected areas forming a coherent ecological unit could ensure the sustainable protection of the Alpine landscape, as well as the continuity of its natural dynamics. They asked the Alpine Network of Protected Areas to analyse the actual potential of protected areas and transboundary links, and to propose some concrete measures. What emerges from this study is that the Alpine region contains several transboundary protected areas, as well as vast protected areas covering over 1,000 ha. It is possible to envisage the possibility of an ecological continuity between these sites, from the Franco-Italian border, to the eastern border of Austria. Out of eight pilot areas, several have been analysed using indicators, and have been recognised as having significant ecological potential to be ecological corridors, or linkage areas. Many protected areas are linked by them, either crossing national borders or within a country. International borders shared between different categories of protected spaces are estimated to cover more than 250 km, and collaboration between these areas could be a driving force behind the establishment of biological connections.

Table 4.3   International conventions mentioned

Date	Name	Place	Scope	Implementation Date
1971	Convention on Wetlands of International Importance especially as Waterfowl Habitat	Ramsar	World	21 December 1973
1972	Convention Concerning the Protection of the World Cultural and Natural Heritage	Paris	World	17 December 1975
1979	Convention on the Conservation of Migratory Species of Wild Animals	Bonn	World	1 November 1983
1979	Convention on the Conservation of European Wildlife and Natural Habitats	Bern	Europe	1 June 1982
1991	Convention on the Protection of the Alps	Salzburg	Alps	6 March 1995
1992	Convention for the Conservation of Biodiversity and the Protection of Priority Wilderness Areas	Managua	Central America	11 January 1995
1992	Convention on Biological Diversity	Rio de Janeiro	World	29 December 1993
1994	Protocol on the Implementation of the Alpine Convention Relating to the Conservation of Nature and the Countryside	Chambéry	Alps	18 December 2002
1995	Agreement on the Conservation of African-Eurasian Migratory Waterbirds	The Hague	Europe	1 November 1999
2000	The Lower Danube Green Corridor Agreement	Bucharest	Danube	5 June 2000
2003	Convention on the Protection and Sustainable Development of the Carpathians	Kiev	Carpathian Mountains	4 January 2006

The section of the study carried out in the area of the Mercantour National Park, the Alpi Maritime Nature Park and the Alta Valle Pesio e Tanaro Nature Park, shows that even this very isolated region does indeed serve as biological corridor (Alpine Network 2004). This has been confirmed by a follow-up study carried out on certain traceable ibexes that, on departing from the Mercantour National Park, moved in a south-westerly direction until they reached the geological reserve of

Haute-Provence. The Alpine network is also associated with other co-operation mechanisms outside of the Alps. A network of protected areas in the Carpathian Mountains is at the planning stage, as is a similar initiative in the Pyrénées. Since these three massifs form an ecological continuum on a macroscopic scale, partnership projects are foreseen.

Conclusion

The conservation of protected areas as a means to protect natural habitats still appears essential in order to guarantee the long term survival and conservation of certain natural environments. On the one hand, the integration of protected areas into ecological networks, leading to their dilution into vaster spaces, does not seem desirable. On the other hand, it is important to emphasise the complementary nature of these two methods of nature conservation. Nevertheless, the functions of protected areas tend to evolve, which modifies the actual status of territories that, from strictly protected and supervised areas, become the project territories or experimental areas of a form of sustainable development destined to expand beyond their boundaries. Integrating the conservation objectives into the sectoral land use planning policies should also create ecological infrastructures, beyond core zones or corridors. However, this zoning can lead us towards mechanisms that reconcile various activities without any clearly defined prioritisation, and it is important, looking beyond the models for the integration of protected areas into ecological networks, to monitor the efficiency of these systems, particularly through an evaluation process that is yet to be defined.

References

Ahern J., 1995 – Greenway as a planning strategy. *Landscape and urban planning*, 33: 131–155.

Bennett G., Wit P., 2001 – *The development and application of ecological networks*. Gland, IUCN, p. 137.

Bonnin M., 2008 – *Les corridors biologiques. Vers un troisième temps de la conservation de la nature*. Paris, L'Harmattan, coll. Droit du patrimoine culturel et naturel, p. 270.

Brunner R., 2002 – *Identification des principales zones protégées transfrontalières en Europe centrale et orientale*. Strasbourg, Éditions du Conseil de l'Europe, coll. Sauvegarde de la nature, p. 26.

Burel F., 2003 – *Landscape ecology. Concepts, methods, and applications*. Plymouth, Science Publishers, p. 362.

Montréal, CBD, 2005 – *Review of experience with ecological networks, corridors and buffer zones*. Programme of Work on Protected Areas, p. 125.

CEDRE (ed.), 2002 – *Le zonage écologique*. Brussels, Bruylant, p. 309.

Cibien C., 2006 – Les réserves de la biosphère: des lieux de collaboration entre chercheurs et gestionnaires en faveur de la biodiversité. *Natures, Sciences, Sociétés*, 14: 84–90.

Fabos J. G., Ryan R. L., 2004 – International greenway planning: an introduction. *Landscape and Urban Planning*, 68: 143–146.

Fall J. J., Egerer H., 2004 – Constructing the Carpathians: the Carpathian Convention and the search for a spatial ideal. *Journal of Alpine Research*, 92 (2): 98–106.

Frolova M., 2000 – Le paysage des géographes russes : l'évolution du regard géographique entre le xixe et le xxe siècle. *Cybergéo*, 143. http://www.cybergeo.eu/index1802.html

Glowka L., Burhenne-Guilmin F., Synge H., 1996 – *Guide de la Convention sur la diversité biologique*. Gland, UICN, coll. Environmental Policy and Law Paper, p. 193.

Jardin M., 1996 – Les réserves de la biosphère. *Revue Juridique de l'Environnement*, 4: 375–385.

Jongman R. H. G., 1998 – Des éléments naturels indispensables. *Naturopa*, 87: 4–5.

Jongman R. H. G., Pungetti G., 2004 – *Ecological networks and greenways. Concept, design, implementation.* Cambridge, Cambridge University Press, coll. Landscape Ecology, p. 345.

Kavaliauskas P., 1996 – "The Nature Frame". *In* Bennett G., Nowicki P. (eds.), *Perspectives on ecological networks*. Tilburg, ECNC: 93–100.

Kiss A., Beurier J.-P., 2004 – *Droit international de l'environnement*. Paris, Pedone, p. 502.

de Klemm C., 1998 – "Voyage à l'intérieur des conventions internationales de protection de la nature". *In* Prieur M., Lambrechts C. (eds.), *Les hommes et l'environnement. Mélanges en hommage à Alexandre Kiss*. Paris, Éditions Frison-Roche: 611–652.

Réseau Alpin, 2004 – *Étude "Réseau écologique transfrontalier"*. Gap, Réseaux alpins des espaces protégés, coll. Signaux alpins, p. 240.

Romi R., 1990 – Convention révolution ou convention inutile ? *Les petites affiches*, 130: 13–17.

Sepp K., Kaasik A., 2002 – *Development of national ecological networks in the baltic countries in the framework of the pan-European Ecological Network*. Gland, UICN, p. 158.

UNESCO, 1996 – *Les réserves de la biosphère. La Stratégie de Séville et le cadre statutaire*. Paris, Unesco, p. 20.

Walmsley A., 2006 – Greenways: multiplying and diversifying in the 21st century. *Landscape and Urban Planning*, 76: 252–290.

Chapter 5

Financing Protected Areas in Madagascar: New Methods

Philippe Méral, Géraldine Froger,
Fano Andriamahefazafy and Ando Rabearisoa

Over the last few years, the problems linked to financing protected areas in developing countries have been the subject of some attention. They illustrate the increasing commodification of nature, the role of intermediary being played by international NGOs, and the emergence of large-scale approaches to conservation; all these having been mentioned in the introduction of this publication.

The inadequacy of traditional financing system for protected areas has been noted during both the 5th World Parks Congress (in Durban in September 2003) and the 7th Conference of the Parties to the Convention on Biological Diversity (in Kuala Lumpur in February 2004). This led to many initiatives, irrespective of the continent or the ecosystem concerned (Emerton et al. 2006).

Indeed, in developing countries, protected areas receive on average 30% of the funding needed to carry out the basic management required for conservation initiatives (Spergel 2001). During the last six years, the governments of many countries, from Africa in particular, have reduced their budgets for protected areas by more than 50%, due to financial and political crises. Several protected areas have become mere 'paper parks' because fundings were insufficient to cover the salaries or vehicles costs, for example.

In addition to public budgets receiving funds through usage fees, taxes and other dues, and to subsidies and donations from international NGOs and aid agencies, other sources of funding such as payments for ecosystem services are on the rise (Emerton et al. 2006; Gutman 2003; Wunder 2005; Pagiola et al. 2005). This new trend results from the fact that industrialised countries find it difficult to increase traditional international aid in a period marked by shrinking public budgets and the criticism of aid efficiency in general. It also results from the development of public-private partnerships against a background of economic globalisation and an increase in direct investment abroad. Finally, it results from the increasing commodification of biodiversity, along with a return to more preservationist policies.

Our objective in this chapter is to illustrate this new trend, and to evaluate its characteristics using the case of Madagascar. More specifically we intend to show how current policy for the expansion of protected areas (Carrière et al., this publication) goes hand in hand with the development of new 'sustainable' financial instruments. Indeed, protected areas are often used as examples of the importance of market mechanisms (trust funds, tourism concessions, etc.). Conversely, the policy of protected areas expansion find in these financial mechanisms an economic legitimacy which, in principle at least, is supposed to justify their existence (Carret and Loyer 2004).

Of all the financing tools available to a state, it seems important to distinguish, on the one hand, those qualified as endogenous, i.e. self-maintained within the country concerned (public financing, taxes and entrance fees among others) and, on the other hand, exogenous, i.e. international sources of funding (trust funds and carbon sequestration projects). Thus, in the first section we highlight the importance of the need for funding, and the problems linked to endogenous financing system. This analysis will reveal why environmental policy actors are aware of the requirement to find alternative sources of funding. Regarding these, we examine in the second section the creation of the Madagascar Foundation for Protected Areas and Biodiversity that aims to manage capital invested in the international stock markets. And finally, in the third section, we examine carbon sequestration projects involving international actors, governments and multinationals.

The Limits of Endogenous Financing Mechanisms

Although the financial sustainability of Malagasy environmental policy has been part of the Environmental Charter since 1990, it was in 2003[1] that its implementation in the more targeted domain of conservation became a priority. Indeed, in September 2003 during the 5th World Parks Congress held in Durban, the President of the Republic of Madagascar, Mr Marc Ravalomanana, pledged to increase, within five years, the surface area of protected areas to 10% of the Malagasy territory. This IUCN standard in fact entails a tripling of this extent from 1.7 million ha to 6 million ha[2].

To date, most Malagasy protected areas have been managed by the National Association for the Management of Protected Areas in Madagascar, now called

---

1   Underlying the environmental policy of Madagascar is a 15 year, three-phase Environmental Action Plan (EAP). Each phase is called an Environmental Programme: EP 1 (1991–1996), EP 2 (1997–2002) and EP 3 (2004–2008). The EAP is based on the concept of the National Environmental Action Plans (NEAP) developed by the World Bank at the beginning of the 1990s, and on the Environmental Charter (Andriamahefazafy and Méral 2004; Chaboud et al. 2007; Froger and Méral 2009).

2   In order to accomplish this work, the 'Durban Vision' group which encompasses the main actors of the environmental policy, conservation NGOs (CI, WCS and WWF), the Department of Water Affairs and Forestry, as well as the main donors, i.e. American, Japanese and French co-operation, intends to complete two tasks: defining new protected areas and classifying them according to IUCN standards (for more details cf. Andriamahefazafy et al. 2007, and, in this publication, Rodary and Milian on IUCN categories, as well as Carrière et al. for a map on protected areas in Madagascar).

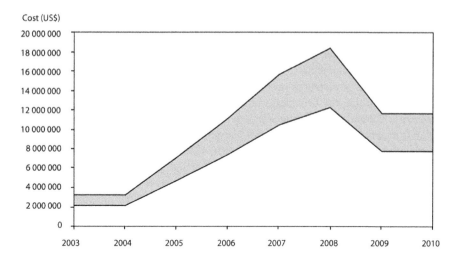

Figure 5.1   Estimated costs for the Malagasy system of protected areas

Madagascar National Parks (MNP). However, those to be created as part of the 'Durban Vision' will be managed from outside the national network, in what is today called Madagascar's System of Protected Areas (SAPM – système d'aires protégées malgache). In fact, this system includes all types of management, whether private, community-based or other. This means that, to achieve the Durban objectives, the modes of governance will have to go through a major evolution; one which, however, cannot be clearly defined at this stage.

This expansion policy makes it even more pertinent for conservation actors to consider methods of funding the SAPM. Indeed, the increasing number of protected areas and their total surface area in particular, entails very high operating costs, as indicated in Figure 5.1 above, which shows an estimate of the total annual financial cost (capital and recurring costs) linked to the increase of protected areas in Madagascar. The creation of protected areas involves costs related to the initial investments. Subsequently, the recurring costs (e.g. salaries, fuel, etc.) must also be taken into account, such that the annual total amount for financing the SAPM was estimated at $15 million in 2007, between $11 million and $18 million in 2008, and between $8 million and $12 million for subsequent years.

Protected Areas cannot be Financed on a State Budget

As a general rule, receiving funding from a government can be more advantageous than from international aid agencies whose programmes do not normally last more than five years, which is of course incompatible with the long term nature of biodiversity conservation. The endogenous dimension of the funding is then understood to be a source of sustainability.

In Madagascar, the situation is unfortunately highly unfavourable towards this type of funding, considering the low availability of public resources. Indeed, since 1990, when the National Environmental Action Plan (NEAP) was established, contributions to finance protected areas in Madagascar came mainly from donors. In this regard, Andriamahefazafy and Méral (2004) have shown that the expenditure of the Malagasy government on protected areas represented less than 2% of the total budget during the first phase of the Environmental Programme (EP 1), and between 15% and 20% during EP 2. It is fair to say that Madagascar benefits today from a network of protected areas largely thanks to donors, of which USAID contributed 68% of the expenditure between 1991 and 1996. As such, the state's capacity to participate in the financing of protected areas has always been low, mainly due to having a small public budget. The Malagasy state, through its own resources and taxes, has very little at its disposal to ensure the implementation of its own environmental policy and, *a fortiori*, the political choices which underlie protected areas.

Moreover, the state also finds it difficult to mobilise more specific fiscal resources, despite the existence of several fiscal projects (on gambling, fuel consumption as well as mineral and oil prospecting) intended to generate revenue for the environment. Even if other special funds paid for by targeted taxation exist in other sectors (e.g. taxes on fuel consumption to fund the Road Maintenance Fund, taxes on alcohol and tobacco to fund the Fund for the Promotion of Youth and Sports, among others), the problems and constraints remain. The tax collection rate is low in Madagascar (around 11%) and the reallocation of taxes in favour of the environment is quite random. It is also not uncommon that these tax revenues find their way into the general budget or the administration. Finally, it is difficult to ignore the lobbying of those economic actors likely to be taxed.

For these reasons, financing through state resources is inconceivable in Madagascar when it comes to the expansion of protected areas. Implicitly, the Durban Vision presupposes the development of new mechanisms.

The Relative Efficiency of Park Entrance Fees

Entrance fees constitute a common source of funding for protected areas. In some cases they generate enough income to cover much of the operating costs of a protected area or a park, particularly when there are many visitors and entrance fees are relatively high[3]. However, many parks set entrance fees way below the amount international visitors would be prepared to pay. The potential for increasing entrance fees is limited for parks that are not well known, or that do

---

3   For example, the Galapagos National Park in Ecuador asks foreign visitors to pay $100 per person as opposed to only $6 per person for Ecuadorians. The number of visitors keeps growing year on year and is currently approaching 80,000. National parks in Kenya, Tanzania, Uganda and Botswana ask foreign tourists to pay a daily fee of between $20 and $30 per person (Spergel 2001).

not contain animal species with a high tourist value. Furthermore, the income stemming from these entrance fees is not always reinvested in the maintenance of the protected areas where it was collected in the first place. This restricts long term self-financing.

It is for this reason that certain countries now allow a protected area to keep a significant share of its entrance fees. Spergel (2001) explains that, in Ecuador, the "special law for the Galapagos" stipulates that 90% of the $100 entrance fee received per visitor must be used for the protection and conservation of the natural environment. Only a few protected areas around the world are able to cover their operating costs through entrance fees alone. Such incomes (and other associated usage fees) must be perceived as a way of complementing, rather than replacing, budget allocations and subsidies received from governments and donors, respectively.

This uneven situation is also found in the case of Madagascar. Promoting ecotourism has always been at the forefront of the strategy for the economic valorisation of biodiversity in Madagascar during the NEAP. It is true that tourism has become one of the country's most dynamic economic sectors. However, the number of tourists remains relatively modest. The highest was in 2005, estimated to be 285,000 visitors. Considering that only around 60% of tourists visit protected areas, and that entrance fees for foreign visitors are between $1 and $5, the income generated by MNP can be estimated at between $171,000 and $855,000 per year. In the best case scenario, MNP can expect to benefit from half of this sum for its own financing, i.e. $427,000 (which represents between 10% and 15% of the operating costs of the institution)[4]. The other half goes towards financing projects intended to benefit the local population, which is not an inconsiderable amount locally, although it appears insignificant in view of the scale of the network.

This situation should deteriorate with the increase in the number of protected areas, which is mechanically going to reduce the impact of ecotourism. First, with the same number of tourists staying for the same amount of time, the number of visitors per park will decrease, along with the income from entrance fees. Second, an increase in the surface area of these protected areas means an increase in the number of *communes* and villages affected. Even with the same amount of visitor entrance fees, the income generated by local rental tourism will be reduced. Only an increase in the number of tourists could mitigate this problem.

---

4   This percentage is fairly high, for it is based on a record number of tourists in Madagascar paying an entrance fee of $5 the highest fee applied in 6 of the 38 MNP sites for a stay of four days or more. This percentage corresponds to the percentage estimated in the United States and Canada a few years ago, for which the share of entrance fees in the recurring costs of their parks was estimated at 17 to 18%. Today, this percentage approaches 34% in the United States, which reveals a strong rate-fixing policy in these countries (Eagles 2001).

From Entrance Fees to Tourism Concession: A Circuit that Excludes
Local Populations

Over the next few years, the financing of protected areas through ecotourism in
Madagascar should result from the implementation of a system of concessions
involving private operators. 2006 saw the beginning of a consideration of the legal
conditions for implementing concessions in certain protected areas, thereby giving
form to the general tendency in favour of the concession market (Karsenty and
Weber 2004).

Concessions concern mainly ecotourism activities in nature parks. Their
principle consists in granting a private body the right to undertake economic
activities (i.e. the right to exploit the tourist accommodation infrastructure, the
shops, the restaurants, etc. as well as the right to set up income-generating visitor
services or excursions), in exchange for fees paid directly to MNP, in accordance
with an amount defined on a case by case basis. Several types of concession already
exist, and differ according to the nature and duration of the contract (long lease,
lease management, management contract, etc.). The financial logic underlying a
concession is that while the management of the economic activity is delegated,
MNP can take part in the general financing of the protected areas.

Concessions offer certain advantages for private operators and especially for
MNP. Indeed, the national association not only sees the possibility of benefiting
from taxes as a source of financing, but also of inheriting the infrastructure created
by the private operator at the end of the contract.

In general, the establishment of concession contracts relies on several laws,
including the Act n°60-004 of 1960 on State-Owned Property, and the Act n°97–
017 of 1997 on the revision of the Forestry Legislation[5]. The Environmental Charter
and the Code of Protected Areas also favour the implementation of concessions
in protected areas in Madagascar[6]. However, the legal dimension also represents
an obstacle, as enforcement decrees specifying the terms and conditions for
implementing concessions are generally very long to be published. Furthermore,
the multitude of laws has subjected the implementation of concessions to several
interpretations, thereby undermining negotiations between contracting parties.

---

5   Act n°2005–019 of 17 November 2005 defines a concession as "the administrative
procedure through which the administration in charge of the public domain grants a natural
person or a public or private legal entity the use of a specific land and the immovables
contained therein, in the conditions determined in the procedure and in the specifications,
for a consideration, for a determined period".

6   Thus, Article 34 of the Code of Protected Areas stipulates that MNP is
authorised "to enter into commercial agreements or otherwise with any natural person or
legal entity and to exercise of its own accord or in partnership, within the framework of the
development of the protected area of the national network or its components, any activities
likely to generate additional income without going against the protection or conservation
objectives".

It is important to note that implementing concessions in certain protected areas currently managed by MNP is part of a more general approach intended to implement concessions in the forests of Madagascar. This approach is clearly targeting protected areas that are not managed by MNP, within the framework of the SAPM, and will be the subject of a concession contract with the Department of Waters Affairs and Forestry, either by tender if the initiative comes from the state, or by mutual agreement if the request is made by a private operator.

While this concession contract policy seems to be in harmony with the establishment of the SAPM, it is not free from the risk of conflict with local populations and/or traditional authorities. Because of the lack of efficient forest management (e.g. through a permanent or regular presence on the ground) by the administration, due mainly to a lack of means, these populations and authorities were able to maintain some of their rights. The private operator will be considered as the sole manager of the area concerned and will only report to the granting authority, i.e. the state[7]. Even if USAID highlights the need to "take into consideration the social aspect and the integration of the local populations into the implementation of the concession policy" (USAID Madagascar 2006: 88) and recommends the implementation of the development plan, it is far from certain that mere 'good intentions' are sufficient.

Moreover, while private operators will be paying fees to MNP, the latter will not be redistributing the money to the *communes* or the local populations (currently receiving 50% of the entrance fees). The advantages for the *communes* and local populations will depend on the side-effects of an increase in the number of tourists visiting the concession sites. This argument is put forward by the promoters of concessions and remains, for the time being, hypothetical. According to Chaboud et al. (2004), Andrianambinina and Froger (2006) who studied the limitations of the aforementioned side-effects, the advantages would have been more direct if part of the fees were allocated to the budget of the *communes* concerned.

Generally, these sources of so-called endogenous financing (entrance fees and concessions) are relatively limited and can only concern a small portion of protected areas, where the potential for tourism is significant. Finally, one can note that these methods of economic valorisation through tourism are constrained by the fact that tourism is part of a global network in which the decision-making power lies in the hands of Western tour operators (Requier-Desjardins 2005). The fact that foreign tourists visit nature parks in Madagascar is the result of competition with other international destinations cashing in on their very own assets. The capacity of the Malagasy state to release enough financial resources to implement the concessions depends mainly on the attractiveness of the concession

---

7   Article 24 of the Draft Decree for the implementation of concessions stipulates that "the private operator is responsible, and at his own expense, for all the precautions required to be carried out professionally to avoid abnormal disruptions to resident properties and local populations".

sites, which in turn depends on the way in which Malagasy tourism differentiates itself from the other destinations of this globalized network.

External Financing: Foundations, Debt-for-Nature Swap and Trust Funds

The types of external financing that have held our attention, are the foundations, the debt-for-nature swaps and the trust funds. The current tendency in Madagascar is to combine these three instruments into one tool for sustainable financing.

The Madagascar Foundation for Protected Areas and Biodiversity

One of the major elements of Madagascar's environmental policy is the creation of a Foundation for Protected Areas and Biodiversity (FAPB) (Andriamahefazafy et al. 2007). One of its objectives would be to facilitate relations with foundations in developed countries in order to attract the donations these foundations are likely to grant for biodiversity conservation initiatives.

As a general rule, these foundations rely at first on debt-for-nature swaps[8]. For aid agencies (and intermediary conservation organisations), these swaps offer a means of financing more conservation initiatives in the debtor country. For the governments of developing countries, these swaps help them to reduce their external debt and to finance projects inside the country. However, these swaps can be complex to realise and very often require the intervention of technical experts from several external institutions, which restricts the autonomy of the governments[9].

As early as 2001, a committee for financial sustainability sought to identify sustainable funding methods for protected areas which could be used in Madagascar. After visiting other countries that had developed similar initiatives (Costa Rica, Mexico and Peru among others), and after examining several options (including green taxes), the committee opted for the establishment of a trust fund[10]. The idea of creating a foundation to manage the trust fund was proposed as early as September 2001. In parallel with work on EP 3 (2002–2003), the

---

8   Debt-for-nature swaps "are a mechanism by which public debt is purchased at a discount by an outside agency – often an international NGO – and retired in exchange for government commitments to fund conservation activities" (Emerton et al. 2006: 46).

9   Madagascar has had many debt-for-nature swaps, and was even the first African country, as early as 1989, to develop a bilateral exchange between the Malagasy government and USAID. It was for the remission of a \$2.1 million debt, with the WWF acting as intermediary (Moye and Paddack 2003).

10   A trust fund is defined as a sum of money or other assets that can only be used to reach one or several specific objectives: financing a single protected area, an entire network of protected areas at the national level, the conservation of a specific species or conservation projects led by local communities and NGOs. It must be separated from other financial sources (such as the regular budget of the governmental agency), while being

donors and the Malagasy government defined a legal, financial and organisational framework for the foundation that relied on the 1995 Foundation Act, and defined four objectives: conservation, research on biodiversity together with ecological monitoring in protected areas, the promotion of ecotourism and environmental education. Finally, following legal audits, the 1995 Foundation Act gave way to a new law in 2004, which increased the autonomy of the foundation vis-à-vis the state and the administration. The interest rate was alleviated, as was the exchange methods between foreign and national income. On the initiative of Conservation International (CI) and the World Wide Fund for Nature (WWF), the Foundation was then created in January 2005.

The functioning of foundations is adapted to the management of the money gathered, either through donations by other foundations or through debt remissions, or in the form of trust funds. As such, trust funds have been completing the current system of financing for protected areas in Madagascar. The research carried out in several countries by Bayon et al. (2001), Emerton et al. (2006) and Spergel (2002) highlights that conservation trust funds can represent sources of sustainable financing for protected areas. Such funds enable the redistribution of significant international subsidies in the form of smaller subsidies, and their extended use over several decades. They are able to reinforce participative governance via the appointment of representatives from NGOs and the private sector, as members of the board committee of the fund, and via the granting of direct subsidies to NGOs and other organisations of the civil society.

Nevertheless, trust funds have several limitations. Administration costs are high, particularly when the capital of the fund is relatively limited or when the fund subsidizes significant technical assistance with designing and implement projects. If the investment strategy, if there is one, has not been well designed, the returns generated by these funds are sometimes relatively low, or unpredictable, especially in the short term. Also, the board committee can be persuaded to finance unrelated projects with no common objectives, if the aims and allocation criteria of a trust fund are not clearly defined within the legal documentation, right from the start.

The financial structure of the FAPB relies on the principle for which the operating costs of MNP are being financed, estimated at $3 million annually for a surface area of 1.2 million ha. The donors have worked out the capital required to obtain annual returns equivalent to the operating costs. This capital is estimated to be $50 million and has become the objective of EP 3[11]. The two founding

---

managed and monitored by an independent board committee. Trust funds can take on one of the following forms: an endowment fund, a sinking fund and/or a revolving fund.

11   Even if the memorandum of association of the foundation mentions that such interests can be used for activities other than those carried out by MNP (i.e. promotion of ecotourism, creation of new protected areas, etc.). This ambiguity is actually a source of misunderstandings between conservation actors since in theory, the objective of the foundation is to finance the SAPM (which includes the protected areas of the MNP's

institutions (CI and WWF) have sought and obtained fairly rapidly agreements in principle from the other donors involved in the NEAP. The foundation began its activities in 2005 with capital of around $5 million (which was valued at $17 million one year later). The money thus invested is used to purchase securities on stock markets, the FAPB having undertaken to respect the compatibility of its investments with its mission, although this cannot be verified.

The governance of such a foundation calls for three essential questions: How will the Foundation allocate non priority funds? Which proportion of MNP's need will the Foundation be able to cover in the end? And finally, does a conflict of interest arise if the Foundation mainly finances activities promoted by its founding members (state, WWF, CI, etc.)? By providing a lobbying function with foreign financiers, by sitting on the board committee of the Foundation and by being field operators and therefore likely to benefit from foundation aid, conservation NGOs *de facto* hold considerable power in this type of institutional set up.

Climate Change and Carbon Funds

With the recent development of payments for ecosystem services, the financing opportunities available to conservation interventions are increasing, linking conservation actors in developing countries directly with international financiers. The case of biodiversity offsets related to climate change and more particularly of carbon sequestration projects intended for reducing greenhouse gas emissions (reforestation, the prevention of deforestation and forestry management among others), reveal this new search for exogenous funding.

Acting Locally, Paying Globally?

In order to fight against climate change, several initiatives have recently become internationalised. Beyond the commitments related to the Kyoto Protocol and the mechanisms created within this framework (joint implementation and Clean Development Mechanism), which enabled industrialised countries to fulfil their commitments to reduce their greenhouse gas emissions, and for accessing the emission permit exchange market, several other mechanisms today make it possible to finance conservation for the countries of the South that are ultimately called upon (when the Kyoto commitments will be renegotiated in 2012) to also take part in the global effort to fight against greenhouse effects: the funds put together by the World Bank (the BioCarbon Fund and the Prototype Carbon Fund),

---

network and any other created outside MNP), when the amount of the funds required for its operating has been calculated on the basis of the recurring costs of MNP.

the national and regional stock markets (the Dutch and European stock markets respectively), as well as the unilateral initiatives of companies[12].

Thus, for a few years now, several countries and private companies have foreseen and developed 'carbon projects' in developing countries, either in isolation or within a regional or national framework. Their objectives are many: facilitating commercial integration with the host country, trying to achieve an ecological and humanistic image, or simply opening up the possibility of negotiating carbon credits. In the last case, the idea is to obtain credits at the lowest cost possible, i.e. where carbon sequestration operations are the easiest.

This is the case for Mitsubishi in the protected area of Makira, in the North-East of Madagascar. The carbon project, initiated by CI, the Wildlife Conservation Society (WCS) and NatSource Japan Co. Ltd (Mitsubishi Group), works on the assumption that sustainable land use helps to protect forest sectors with a high biological diversity. The investments of this firm then cover a portion of the management costs of the protected area. The proximity of international institutions (international conservation NGOs as well as the World Bank, the Global Environment Facility (GEF) and bilateral co-operations) able to act as an interface between developing countries and multinationals, facilitates this type of initiative[13].

This funding method has the advantage of accelerating the granting of funds for local associations, which supports activities of economic valorisation by establishing production networks (craft industry, apiculture, etc.). It also has the advantage of ensuring a continuous flow of income over the long term, in which case the farming groups can choose the best way to deploy these resources.

---

12   For the time being, the Clean Development Mechanism (a flexible mechanism stemming from the Kyoto Protocol) does not allow to finance actions to prevent deforestation, which could be assimilated to the direct financing of protected areas. However, many NGOs manage to finance protected areas by resorting to reforestation projects within the same protected areas. In this case we are dealing with bilateral arrangements or, more simply, with conditional aid intended to have an interest in conservation, rather than participating in a hypothetical international carbon market.

13   Since 2005 several projects have thus been developed in Madagascar, in or outside protected areas, such as the Mantadia-Zahamena Biodiversity Conservation and Restoration Corridor Carbon Project, or the Makira Carbon Fund in the region of Maroantsetra. The process is as follows: the sponsor identifies, through a firm of consultants, the pre-feasibility of a carbon offset project set up in a specific location (identification of the nature of the project, i.e. reforestation, preventing deforestation, etc., and of the actors involved locally, then an approximate estimate of the project costs). If this phase is concluded successfully, then a feasibility study is carried out to refine the project further and work towards meeting the conditions for obtaining certified carbon credits (proving the augmentation of the project, identifying economic leaks and the cost per tonne of carbon, among others). If the project is finally deemed viable, it falls to the various actors to then find potential investors.

From Project to Market Logic

Considering experiences with economic valorisation, it appears that the arrival of a sum of money can destabilise a village community by creating tensions related to the improper solicitation of funds. This can create conflict between the local population and the decentralised administration, particularly the Department of Water Affairs and Forestry. Confronted with such an income, the issues around the delegation of land management and its security can be exacerbated if the project does not take these factors into account.

Moreover, the risk of a market logic developing outside the project, with on the one hand credit salesmen (local farmers' organisations and their related local or national associations), and on the other hand credit buyers (multinationals or brokerage companies), is far from negligible. The decision whether or not to buy carbon credits should logically relate to a comparison between prices in the emission permits market, the potential for technological improvements in the production processes of firms internally, and the costs of carbon credits offered by other projects, such as energy projects for example. The local communities might fail to master the process, since their projects will be in competition with other projects in other countries and, more globally, with the different options being offered to firms. There is a risk that many offers will not find a buyer or, in the best case scenario, will find a buyer via a process that will escape them altogether. As long as the situation is underpinned by pilot projects led by pioneering companies, these risks will be reduced. However, as carbon markets expand, investors should become mere credit buyers, thus leaving project initiators to cover the risks (Conservation Finance Alliance 2003).

This situation, and the future risks are not wholly perceived by the actors of the environmental policy of Madagascar. It is essential to find short term funding methods, since the socioeconomic risks of future projects are not taken into consideration in either the projects or the discourses. The vulnerability of farmers' organisations in such a market system is of no real consequence. By being at the interface between intermediary associations and/or the rural communities on the one hand, and the multinationals on the other, the most influential conservation NGOs take on the function of a financial intermediary, a function that, albeit new, mobilises their lobbying activities with private financiers and their role of historic intermediary in the local landscape.

The emergence of these new financing opportunities is compatible with policies for the expansion of protected areas. For many authors, the existence of transaction costs with a not insignificant fixed percentage encourages the promotion of larger scale projects (Wunder 2005; Pagiola et al. 2005). As highlighted by Smith and Scherr (2002: 31), "the bigger the area, the more tons of carbon involved and the lower the unit costs of items like project design, management and certification". The development of market mechanisms with an international dimension, and the sequestration of carbon or others, is likely to promote far-reaching projects. Conversely, it will be more difficult to appropriate these projects locally (for example, the difficulty in co-ordinating farmers' organisations over an important number of villages or *communes*). The problem encountered by NGOs promoting projects with a view to obtaining payments for ecosystem services over large surface areas, is to find associations, at the local level, that can take over: associations that have sufficient influence towards the local population, and that have at their disposal adequate internal structures (personnel, operating means, etc.) to cover fairly large surface areas. Finally, by shifting the decision centre to the international level, i.e. in an environment made up of international NGOs and multinationals, there is a great risk that these projects will be disconnected from the local situation on the ground.

Conclusion

Thanks to new financial instruments, the policy for the expansion of protected areas in Madagascar gives an economic legitimacy to its sustainable financing objective. The fact that, today, donors associate the financial sustainability of an environmental policy with the sustainable financing of protected areas, is a good reflection of this shift. Conversely, NGOs and donors that promote these instruments rely on protected areas to justify the economic interest of conservation and the use of these instruments.

The Malagasy experience shows how, in the space of a few years, the discourse surrounding these new financing instruments has become common to all environmental policy actors. It goes hand in hand with the increase in the surface area of protected areas in Madagascar, without necessarily guaranteeing efficient financing.

International financial instruments contribute to an increase in the number of intermediaries and, in so doing, displace the centres of decision-making and negotiation to centres outside the country (stock markets within the framework of trust funds, carbon markets concerning the tools linked to the negotiations on climate, etc.). This increases the power of conservation NGOs, which can serve as financial intermediaries between foreign sources of funding and park managers, and potentially farming groups. Moreover, this tendency to develop instruments for supplanting state power, assuming they become efficient, runs the risk of straining relations between the decentralised services of the forestry administration and the other environmental policy actors. This problematic goes beyond the Malagasy case, since in many other countries (see the classic case of Costa Rica) the tendency towards the development of these financing instruments is real. Even if the economic and institutional characteristics differ from one country to another, the issues found in the Malagasy case undeniably have an international impact.

References

Andriamahefazafy F., Méral P., 2004 – La mise en œuvre des plans nationaux d'action environnementale : un renouveau des pratiques des bailleurs de fonds? *Mondes en Développement*, 32 (127): 27–42.

Andriamahefazafy F., Méral P., Rakotoarijoana J. R., 2007 – "La planification environnementale : un outil pour le développement durable ?". *In* Chaboud C., Froger G., Méral P. (eds.), *Madagascar face aux enjeux du développement durable. Des politiques environnementales à l'action collective locale.* Paris, Karthala: 23–49.

Andrianambinina D., Froger G., 2006 – "L'écotourisme, facteur de développement durable dans un contexte de mondialisation ? Le cas de Madagascar". *In* Froger G. (ed.), *La mondialisation contre le développement durable?* Brussels, PIE Peter Lang: 281–310.

Bayon R., Deere C., Norris R., Smith S., 2001 – *Environmental funds. Lessons learned and future prospects.* Document, p. 26. http://biodiversityeconomics. org.

Carret J.–C., Loyer D., 2004 – *Comment financer durablement les aires protégées à Madagascar. Apport de l'analyse économique ?* AFD, Notes et documents 2003–2004, n° 4, p. 45.

Chaboud C., Méral P., Andrianambinina D., 2004 – Le modèle vertueux de l'écotourisme : mythe ou réalité ? L'exemple d'Anakao et Ifaty-Mangily à Madagascar. *Mondes en Développement*, 32 (125): 11–32.

Chaboud C., Froger G., Méral P. (eds.), 2007 – *Madagascar face aux enjeux du développement durable. Des politiques environnementales à l'action collective locale.* Paris, Karthala, p.109.

Conservation Finance Alliance, 2003 – *Guide des mécanismes financiers de conservation.* http://www.conservationfinance.org, p. 439.

Eagles P. F. J., 2001 – *International trends in park tourism.* Document prepared for Europarc 2001, Hohe Tauern National Park, Matrei, Austria, 3–7 October, p. 44. http://www.ahs.uwaterloo.ca/rec/research/eagles.html.

Emerton L., Bishop J., Thomas L., 2006 – *Sustainable financing of protected areas. A global review of challenges and options.* Gland/Cambridge, UICN, p. 97.

Froger G., Méral P., 2009 – "Les temps de la politique environnementale à Madagascar: entre continuité et bifurcations". *In* Froger G, Géronimi V., Méral P., Schembri P. (eds.), *Diversité des politiques de développement durable. Temporalités et durabilités en conflit à Madagascar, au Mali, au Mexique.* Paris, Karthala: 45–67.

Gutman P., 2003 – *From goodwill to payments for environmental services. A survey of financing options for sustainable natural resource management in developing countries.* Washington, WWF, Macroeconomics for Sustainable Development Program Office, p. 148. http://biodiversityeconomics.org.

Karsenty A., Weber J., 2004 – Les marchés de droits pour la gestion de l'environnement : introduction générale. *Revue Tiers Monde,* 45 (177): 7–27.

Moye M., Paddack J.-P., 2003 – *Madagascar's experience with swapping debt for the environment. Debt-for-nature swaps and heavily indebted poor country (HIPC) debt relief.* Working document for the Ve World Parks Congress, Washington, Center for Conservation Finance/WWF, p. 19.

Pagiola S., Arcenas A., Platais G., 2005 – Can payments for environmental services help reduce poverty? An exploration of the issues and the evidence to date from Latin America. *World Development*, 33 (2): 237–253.

Requier-Desjardins D., 2005 – La valorisation économique de la biodiversité : ancrage territorial et gouvernance des filières. *Revue Liaison Énergie-Francophonie*, 66–67: 77–81.

Smith J., Scherr S. J., 2002 – *Forest carbon and local livelihoods. Assessment of opportunities and policy recommendations*. CIFOR Occasional Paper n° 37, p. 47. http://www.cifor.cgiar.org.

Spergel B., 2001 – *Raising revenues for protected areas*. Washington, Center for Conservation Finance, WWF, p. 33.

Spergel B., 2002 – "Financing protected areas". *In* Terborgh J., Van Schaik C., Davenport L., Rao M. (eds.), *Making parks work. Strategies for preserving tropical nature*. Washington, Island Press: 364–382.

USAID Madagascar, 2006 – *Politique de mise en concession dans le domaine forestier national incluant le système des aires protégées malagasy*. Document de travail, BAMEX, MCI, p. 88.

Wunder S., 2005 – *Payments for environmental services: some nuts and bolts*. CIFOR Occasional Paper n°42, p. 24. http://www.cifor.cgiar.org.

PART III
New Conservation Territories

New conservation territories are currently taking shape through several dynamics: social and economic interplay, power relations between the inside and outside of protected areas, and the definition of legal frameworks in policies relating to large natural infrastructure and sustainable development. What opportunities do these new territories offer populations who are supposed to make the most of sustainable development areas? Is their size suitable for the social and spatial management of the populations concerned? These questions form the basis of the third part of this publication.

The Guyana Amazonian Park, created in February 2007, is the perfect illustration of the 2006 French legislative reform on national parks. Catherine Aubertin and Geoffroy Filoche recall the painstaking creation of this park, and question the possibilities of interacting with the environment which Amerindian populations have been granted, for lack of indigenous community status. Will these populations be able to develop their way of life between the cores of the park and the zones of free adherence? The creation of a protected area unfailingly entails a redistribution of power. In this regard, it is between the global and the local, i.e. between the geopolitical choices of the state and the indigenous populations, that the local authorities will assert their power.

'Indigenous Lands' represent a specific category among the many protected areas of the Brazilian Amazon. Their legal and ecological particularities, mainly forest areas held in usufruct exclusively by the Amerindian populations, and their importance in terms of surface area, explain many of the challenges and threats affecting them today. Indigenous Lands are preferred targets for local and international NGOs as well as public policies claiming a 'socio-environmental' development model, where such a model is environment-friendly and respects cultural diversity, while defending the maintenance or adoption of methods for the sustainable management of natural resources, concerning areas over which inhabitants have collective territorial rights. As such, Indigenous Lands have emerged as preferred areas of conservation and sustainable development, while their legal existence is not based on the conservation of ecosystems. In this light, Bruce Albert, Pascale de Robert, Anne-Élisabeth Laques and François-Michel Le Tourneau discuss the impact which indigenous territory delimitation in protected areas and the establishment of development experiences have on the local methods of space and resource exploitation management.

Taking into account local populations, protected areas are indeed sometimes used for regional economic activities contradicting the law. Following the recent geographic evolution of pastoralism in West Africa, many protected areas actually represent potential pastoral areas. Jean Boutrais points out several spatial configurations: from protected areas excluding pastoralists to those included in pastoral lands, via intrusion areas under exceptional circumstances. Although they are attractive in view of their biological wealth, protected areas are also dangerous because of the higher health risk they represent for cattle herds. In order to dilute these edge effects, would it not be preferable to integrate conservation areas into pastoralist areas? To date, the only long term experimentations are found in East Africa. They show a propensity for the development of territorial arrangements between natural and pastoral spaces. In fact, this reflects the main issue concerning conciliation between sustainable development and conservation.

The last three chapters of this publication illustrate the current state of protected areas: more than a mere participation of 'local populations' becoming subservient to the positive or negative dynamics of conservation policies, today's issues concern the relations between the world of conservation on the one hand, and politically as well as economically important social groups on the other (e.g. local authorities, Amerindians or pastoralists), where the coherence between these relations and conservation only partially match up. In this sense, the new territories of conservation are more than a bottom-up policy advocated for a long time by conservationists and social scientists; they establish interlocking policies between different social groups, redefining hierarchy of scale and power. Through the different cases presented in the third part of this publication, we can discern a readjustment of participative policies, shifting from a vertical to a horizontal approach and capable of linking previously divergent logics. Only on this ground does it appear that new territories can be built more on co-operation than on dependent relations.

Chapter 6

Creation of the Guyana Amazonian Park. Redistribution of Powers, Local Embodiment and Territorial Divisions

Catherine Aubertin and Geoffroy Filoche

The 2006 legislative reform concerning French National Parks took note of the new conceptions of sustainable development. These allow for the presence of local populations in protected areas and promote their participation in the management of the natural heritage[1]. However, these conceptions underpin projects that are so diverse that they do not always entail a clear break with the prevailing centralised and protectionist tradition. The French reform on National Parks questions the limitations of the human occupation of protected areas, and how local populations will be associated with management measures.

We examine these issues using the example of the Guyana Amazonian Park which was created in February 2007, following the new legislation on National Parks[2]. We recount at first how the international movement for the integration of conservation and development objectives resulted locally in the creation of the park. We then analyse the extent to which the new legislative framework, stemming from many difficult consultations with Guyanese civil society, offers new opportunities to local populations in terms of status, usage rights and territory delimitation. The idea is to examine, based on the analysis of the Decree on the Creation of the National Park, the extent to which this decree will have an impact on the 'traditional' environmental and economic practices of the communities and, in parallel, the extent to which these communities can be prevented from developing their way of life. More specifically, we analyse the process through which the 'local' is embodied and expressed in relation to the central government. In other words, we propose to evaluate how and by whom exactly the Amerindian

---

1   This chapter falls within the framework of the trans-departemental incentive "Protected Areas" of the IRD and within the framework of Evaluating Effectiveness of Participatory Approaches in Protected Areas (EEPA) research programme – IUED/UICN/ MAB/IRD). We would like to thank Françoise Grenand for her comments and careful proofreading of this text.

2   Act n°2006–436 of 14 April 2006 relating to National parks, marine natural parks and regional natural parks. Decree n°2007–266 of 27 February 2007 creating the National Park called "Guyana Amazonian Park".

and Maroon[3] communities are represented, as their interests promoted both through their own traditional institutions, and through the territorial authorities stemming from the common law on decentralisation (i.e. *communes*, Departmental Council and Regional Council) which have more opportunities to capture the decisional levels created by the park.

A Forced Reform: From the Earth Summit to the Giran Report

The creation of the Guyana Amazonian Park is the result of major international environmental conventions and geopolitical issues, involving the presence of Europeans in the Amazon. It was in 1992, during the Earth Summit held in Rio, that former French president François Mitterrand announced the French contribution to the Convention on Biological Diversity (CBD): the creation of a large National Park with Surinam bordering it to the west and Brazil to the east and south (See Plate 12).

Trying to establish a National Park in a French overseas territory is not without difficulty, and the Guyana Amazonian Park was no exception in this case: local officials did not always adhere to the unwelcome directives of metropolitan France. While the creation of a National Park seemed indeed to clash with the legacy of decentralisation, it also brought to light the internal dissensions of the very heterogeneous Guyanese society.

Two projects were initiated successively and abandoned following heavy controversies until, at the Summit on Sustainable Development held in Johannesburg in 2002, former French President Jacques Chirac revived the National Park of Guyana as one of the major works of his presidency. This was undoubtedly galvanised by the declaration of Brazilian President Fernando Henrique Cardoso concerning the creation of one of the largest parks in the world, the Parque nacional das montanhas do Tumucumaque, covering an area of 38,000 km² in the states of Amapa and Para, on the border of the French territory (Fleury and Karpe 2006; Grenand et al. 2006). The Park project, in line with international co-operation as regards protected areas, was the outcome of the development of a treaty peculiar to the Amazonian Basin. The Amazon Co-operation Treaty was signed in 1978 by the Amazon countries, when environmental issues were not on the agenda and when most signatory countries were ruled by dictatorships. In 1998, the Treaty adopted a new image and became the Amazon Co-operation Treaty Organisation (ACTO). French Guyana could not be asked to sign the Treaty, since it is not a state in its own right but the region of a European country. However in 1994, the Executive Secretary of the ACTO as well as the Brazilian Minister of Foreign Affairs, proposed that French Guyana be accepted as an observer in addition to the eight Amazonian countries involved. In this regard,

---

3   Descendants of fugitive Black slaves of the 18th century, formerly called Bush Negroes.

ACTO's links with the European Union would probably open up new markets to the South American region.

In this international and national context, the 1960 French Act on National Parks, which reflected the notion that a park must be protected against all human action, and that its management is the exclusive business of the central state, was significantly reformed. Two main reasons were behind this change. First, this concept of park management was not in line with the new objectives and management methods of protected areas. In accordance with the notion of sustainable development, environmental conservation must serve local populations and be carried out with their participation. In addition, the conservation objectives can only be reached if the local populations can draw benefits from the existence of the park. Secondly, France had become decentralised (a process initiated by the legislations of 1982–1983) and, as a result, had created many new local governments. French Guyana was endowed with a Departmental Council and a Regional Council with their own jurisdiction. French deputy Jean-Pierre Giran was tasked with visiting Guyana to investigate the elaboration of the policy on National Parks in matters of territoriality, decentralisation and international co-operation. This visit was to be decisive in this regard. The new Act of 2006, which is based to a large extent on Giran's report (2003), dedicates the entire second chapter to the "Amazonian Park in Guyana", and ratifies specific regulations aimed at making the creation of the park politically acceptable for the local governments, local populations and NGOs.

The new Act brought major changes for the normative architecture of a National Park. Indeed, today a National Park is endowed with a dual legal system originating in very distinct logics. On the one hand, the absolute central protection area becomes the 'core' of the park, which is the territory of maximum protection. This area is subject to a legal system first established by law (determined by the Parliament) and then specified by the Decree on the Creation of the Park (falling within the competence of the central executive power). It is then specified further by a specific Park Charter resulting from negotiations between the state, local governments, traditional authorities, scientific and institutional key players as well as key players from associations. The peripheral area, on the other hand, has a different logic which is not, like before, determined unilaterally by the central state. This 'Zone of Free Adherence' (ZFA) is a zone of 'sustainable development' to which *communes* decide to adhere by adopting the Charter. It is delimited by scientific (geographic continuity or ecological interdependence with the core area) and political (the political will of *communes* to adhere to the Park Charter concerning all or part of their territory) criteria.

The demarcation of the core area and the potential ZFA emerges from the Decree on the Creation of the National Park. The Charter which is currently being negotiated in Guyana, must be adopted within five years of the creation of the park, in this case 2012. It contains two sets of standards: the first set specifies what should be recorded in the legislation and the Decree, as regards the core of the park. The second set of standards establishes a local territory project concerning the ZFA. This has resulted in a complex arrangement for the assignment of jurisdictions and the negotiation of spaces (See Box 6.1).

The success of the Guyana Amazonian Park will depend on how the different interests of the local populations, who had been fairly uninvolved until the establishment of the park, will converge in time. What guaranties and constraints can the legal framework offer these populations?

Box 6.1   The institution of the Guyana Amazonian Park

The Charter proposal is elaborated by the Park Board. The Park Board is made up of a Park Director, a Board of Directors, an Economic and Social Committee (called "Committee for Local Life" in Guyana) and a Scientific Committee. On the Board of Directors, local representation is made up of 12 members from the local government and 5 members from the traditional authorities, which represents a majority compared to the state representation that only includes 10 members.

The Zone of Free Adherence (ZFA) is governed by common law, i.e. as if the Park did not exist, which is made more coherent by the Charter from the viewpoint of the sustainable development project. Thus, it is possible to modify the common law (e.g. town planning legislation) to this end, by following the classic procedures.

The core areas of the Park (the Guyana Park has three) is governed by formal standards which result from the five-level centre of decisions, from the most general to the most specific:

The Orders of the Park Director implement the principles established by the Charter or depart from it in certain conditions.

These Orders most often require regulations elaborated by the Board of Directors and making explicit the regulations resulting from the Charter.

The Charter clarifies the regulations in force in the Park.

These regulations are laid down by the Decree on the Creation of the National Park. The Decree must observe the general standards included in the legislation of 2006. As it stands, the Park Board benefits from a certain amount of leeway to elaborate and implement new prescriptions. In time, this leeway will be redefined with the draft of new texts and new interpretations (Filoche 2007b).

The Ambiguous Status of the Park's Populations

The participation of the 'local populations' has become a prerequisite for the success of conservation. How should these populations be defined, considering their heterogeneity, and considering that the notion of 'local' is not restricted to the borders of the park? Should local populations benefit from a unique status? In the case of Guyana, while the Amerindians and the Maroons seem to be the main affected parties, they do not represent the local populations in their entirety and, for this reason, they are not the only populations having a particular role to play in the participative process.

The relations between France and indigenous populations provoked many international denunciations regarding human rights and biodiversity conservation issues[4]. How can France recognise the presence of Amerindian and Maroon populations but not their status as indigenous people (specifically in the case of the Amerindians), as traditional populations or ethnically differentiated communities? Does the creation of the Guyana Amazonian Park bring new identity or territorial rights to these populations?

French Nation and Ethnic Specificity

The population of the three core areas and the potential ZFAs are made up of around 7,000 people living in an area of 34,000 km² and distributed within five *communes*: Camopi, Maripasoula, Papaïchton, Saint-Élie and Saül (See Plate 12). Thus, less than 5% of the population of Guyana lives in more than one third of its territory. Those concerned are a small group of Creoles living predominantly in Saül, three Amerindian ethnic groups (Wayãpi, Teko or formerly Emerillon, and Wayana) and Maroon communities (called Boni or Aluku). These populations practice subsistence activities relying on slash-and-burn agriculture, a technique which is well-adapted to the environment, even if land and demographic pressures can damage the viability of this form of agriculture (Renoux et al. 2003). Maroon populations sometimes also practice gold panning, following the example of the many illegal immigrants (the Brazilian *garimpeiros* in particular) whose activities create important public health and security issues, as well as problems related to the degradation of the environment (Collective 2005). These Amerindian and Maroon populations are undergoing painful transformations imposed upon them by modern society. In this context, they view the park either as protective or threatening.

In Guyana, the French government has always refused to legally and politically acknowledge that individuals can be French as well as members of another community constituting a framework of sociability and constraint (Grenand and Grenand 2005), although it did not refuse to do so in Mayotte or New Caledonia. This means that the French government is rejecting the notion of indigenousness, with all that it entails. Consequently, when it is applied to Amerindians and Maroons from Guyana, French law – from the time of the 1987 Decree acknowledging that they have collective usage rights on specific zones[5] – uses the following

---

4   In this regard, the French government did not adopt Convention n°169 of the International Labour Organisation and made a reservation concerning Article 27 of the International Pact relating to civil and political rights, among others. Convention n°169 concerning indigenous and tribal peoples in independent countries, adopted in 1989 and implemented in 1991, was ratified by 15 states. This text advocates the maintenance and development of indigenous peoples as distinct communities within the framework of the states where they live today.

5   Decree n°87–267 of 14 April 1987 on "the modification of the Code of State-Owned Property and relating to state concessions and other acts passed by the state in

circumlocution: "communities of inhabitants who traditionally draw their means of subsistence from the forest". However, recently, Article 33 of the French blueprint law for overseas territories (2000) accepted the terms of Article 8j of the Convention on Biological Diversity, establishing the links between biological and cultural diversity: "The state and the local governments encourage the respect, protection and maintenance of the knowledge, innovations and practices of indigenous and local communities embodying traditional lifestyles relevant for the conservation and sustainable use of biological diversity". Yet, with regard to French law, this is ambiguous or even unconstitutional. In international law, the category of 'indigenous communities' makes it possible to acknowledge territorial rights founded on established occupations and ethnic status. Accordingly, all the states of the Amazon Basin explicitly acknowledge the legal status of indigenous and local communities. However, while Article 33 has not been challenged for the time being, no real legal consequence in terms of identity or territorial rights has been derived from the entry of this controversial category into French law. Even if one can say that the law applicable in Guyana acknowledges that indigenous and local communities have their "own legal existence" (Karpe 2007) and configures a sort of legal status *sui generis*, it is difficult to know accurately and concretely how these communities will fare in terms of territorial management and development project implementation.

Thus, irrespective of whether or not a park is established, local populations are only defined through the notions of 'way of life' and 'usage rights'. While the populations situated on the territories of the *communes* of Camopi, Maripasoula and Papaïchton are known for their knowledge and respect of the forest and its ecology, they still need to be identified by the Charter. Their recognition is subject to the opinion of the traditional authorities serving on the board of directors (Untermaier 2008). In this light, the Charter will determine their official existence and whether they have a different status from the Creole residents' status.

In Brazil, things are very different. The Amerindians and the Quilombolas (descendants of fugitive slaves with a history close to that of the Maroons in Guyana), were acknowledged in the 1988 Constitution. The Amerindians were recognised on the basis of their established occupation of the territory, while the Quilombolas were recognised on an ethnic basis. Both groups were recognised as social groups with rights, regardless of any environmental consideration. It is even acknowledged that they should benefit from a certain autonomy. Only much later was the legal category of "traditional peoples and communities" defined and its usage extended by various decrees for the purpose of sustainable development policies. This category made it possible to reinforce community-based management systems, and to promote the territorial claims of very heterogeneous groups with no reference to their established territorial occupation or their ethnic origin. Rather, they were distinguished according to a common social history and

---

Guyana with a view to exploiting or ceding state property", *Journal Officiel* of 16 April 1987, p. 4316.

a sustainable method of resource appropriation and management. This is mainly the case of the rubber tappers, Brazil nut gatherers and babaçú breakers, but also of the communities of fishermen and river residents (cf. also Albert et al., this publication).

An Official Specificity after All

With the Act of 2006, the French government acknowledged that the parks could have resident populations. It also recognised that certain categories of people could circumvent at least some of the more comprehensive environmental protection measures applicable to the core area. These categories refer to "permanent residents in the core of the park", "natural persons or legal entities exercising permanently or seasonally an agricultural, pastoral or forestry activity in the core", and "natural persons exercising a professional activity on the date of creation of the national park, duly authorised by the Park Board". Residents fitting these categories are granted the right to carry out livelihood activities with fewer constraints, and therefore to ensure that they can live under normal conditions while fully enjoying their rights. However, these activities must also be "compatible with the protection objectives of the core of the National Park".

Do the Amerindian and Maroon communities of Guyana benefit from different advantages? According to the law, "especially considering the particularities of Guyana", it is possible for the Decree and the Charter to make more favourable provisions for the three categories of persons redefined for the Guyanese case. The situation of a "community of inhabitants who traditionally draw their means of subsistence from the forest, for whom collective usage rights are recognised for hunting, fishing and any activity required for their subsistence", is not fundamentally different from that of a permanent resident (e.g. a Creole whose residence is in the core area) or that of a natural person or legal entity exercising an economic activity (e.g. a forestry business). The fact that one can, in the core area, depart from the general environmental protection rules, remains a *possibility* and not an *obligation* imposed by law, which should be respected within the classification Decree and the Charter.

The status of the local communities is not entirely ratified by law. It is up to the Decree on the Creation of the National Park and the Charter to transform this possibility of derogation into obligation, i.e. to explicitly guarantee local communities usage rights in the core area. However, the mission of the Park Board is "to contribute to the development [of these communities], by taking into account their traditional way of life". From this statement, we can infer that the Park Board could indeed apply preferential treatment to these communities of inhabitants. To what extent does the Decree allow such a treatment?

Perspectives for the Development of Local Communities

Without any reference to indigenousness or ethnicity, the park established the special status of local populations, be they Amerindian, Maroon or even Creole. However, it is difficult to determine whether it consolidated certain rights already acquired, whether it enabled the sustainability of environmental practices and whether this led to new development possibilities.

Strengthening the Usage Rights of Local Communities

The French government has for some time acknowledged the presence of Amerindian or Maroon communities, while abstaining from instituting any substantive reforms in terms of granting rights over the land, as well as defining the actual legal status of these communities. Because the forest in which these communities live is part of the private domain of the state, it falls to the central government – and not to the Departmental Council of Guyana – to recognise the usage rights of these communities and the concession of state-owned lands for their benefit, despite the many requests of the Council for the retrocession of lands to the local government.

The Decree of 1987, as previously mentioned, determines the procedure for establishing the 'collective usage rights' on the state-owned lands of Guyana. These rights concern "hunting, fishing and, more generally, exercising any activity required for the subsistence" of the communities of inhabitants which traditionally draw their means of subsistence from the forest. To this end, Zones of Collective Usage Rights (ZCURs) are granted by Order of the *Préfet*. Each Order determines their location, surface area and recipient community. The total surface area of the ZCURs in the park is 5,628 km² and covers a five-kilometre area on either side of all main rivers and tributaries (See Plate 12).

The same decree holds that the Amerindian communities, constituted into associations or companies, can request to benefit freely from a 10-year concession. Such a concession holds that they can utilise state-owned lands situated within a determined area for cultivation, farming or simply for the housing of their members[6]. Since the aim of this decree was to favour settlement above all, theoretically hunting and fishing activities are not authorised in these concessions. It is the *Préfet* who pronounces the definitive or partial withdrawal of the concession, when the members of the association or company have ceased to permanently reside in a given area (although "permanent residence" has yet to be defined), or when the community finds it impossible to fulfil its obligations as defined in the concession (e.g. the land was not developed).

---

6   In the common law system (Article R. 170–38 of the Code of State-Owned Property) concessions are only granted "to a person of age entitled to stay regularly and permanently in Guyana, the concession being granted in a personal capacity". This is a remarkable exception in favour of the communities.

In both the ZCUR and concession cases, the legal situation of the Amerindians and the Maroons of French Guyana in relation to their land is very precarious, since their rights depend entirely on the *Préfet*. No objective condition, such as proof of long-term occupancy or the respect of all the conditions imposed by the state, can ensure the territorial permanence of the local communities, unlike what other states of the Amazon Basin have been doing (Filoche 2007a).

Does the park offer more guaranties concerning the rights to access resources and to maintain usage of the land? The core area is generally a profoundly regulated space: Article 10 of the Decree on the Creation of the National Park specifies that agricultural, pastoral or forestry activities in the core of the park are subject to the authorisation of the Park Director. Moreover, hunting and fishing there are strictly prohibited. Indigenous communities as well as permanent residents are not, however, entirely subject to these provisions.

Indeed, the communities inhabiting the park could benefit from the geographic as well as material expansion of their activities. These communities have rights on the entire core area and not just on zones strictly defined by an Order of the *Préfet* or a concession. They are not subject to the regulations as regards building works or the creation and maintenance of new villages for their own use. They can hunt, fish and practice "traditional slash-and-burn agriculture" freely. They can also remove or destroy non-cultivated plants to build traditional houses, open forest tracks or clearings and make fires (Article 22)[7]. They can even sell off their surplus catch from hunting and fishing exclusively to other members of the communities of inhabitants, or to residents of the park, and vice-versa. A restricted commercial circuit is made possible inside the core area, provided no meat or fish is sold outside the park or to people coming from outside[8]. Furthermore, one can deduce from the Decree that the usage rights granted to the communities in the core zone are more extensive than those granted in the former ZCUR. Tolerated activities are not limited to 'subsistence' activities but also include the craft industry.

Concerning the permanent residents, particularly the Creoles, hunting and fishing must only be carried out occasionally in the core of the park. However, nothing confirms that the collective usage rights of the communities prevail over those of the residents. It is probably the Charter that will determine how to concretely settle potential conflicts over rights between communities and permanent residents around these resources.

Certain crucial questions currently remain unanswered. Collective usage rights applicable in the core area do not have specific recipients, as opposed to the ZCURs allocated to designated communities. The question remains as to how the various communities are going to arbitrate their potential conflicts. Furthermore,

---

7   However, food gathering for selling purposes and even for subsistence feeding, is not mentioned, which is a surprising omission.

8   This prohibition refers to a restrictive definition of what 'subsistence' can represent. Thus, theoretically, the communities will not be able to sell meals to tourists when the basic ingredients of such meals come from the core area.

former ZCURs will be fragmented between the core zone and the ZFA of the park, which will create some confusion, all the more since the Charter could impose limitations on activities performed in the ZCURs located in the ZFA.

Unclear Development Perspectives

The creation of the park must ensure that the populations benefit from conditions for economic development that will respect biodiversity conservation. Do the constraints applicable in the core area allow this type of development?

Under the Act of 2006, the prohibition of industrial and mining activities in the core zone of any park is clearly defined and final. Yet, the park currently has approximately 10,000 illegal gold panners operating within it. This raises questions concerning how they will be removed and how the ban on gold panning will be policed.

The Decree on the Creation of the National Park in Guyana holds that, in general, commercial and artisanal activities are forbidden in the core area, except as we saw, for the communities of inhabitants which, contrary to the permanent residents, can freely exercise artisanal activities. Within this framework, these communities are also able to remove rocks, minerals, non-cultivated plants and non-domesticated animals. However, some ambiguity remains concerning the commercial nature of this activity. For example, the Decree does not prevent communities from selling their craft to people from outside the park. This was against the Park Board's will which had nonetheless been expressed to the drafters of the Decree.

The status and surface area of the ZFA were still being negotiated in 2011. According to the draft project, certain subsistence activities such as gathering, cultivation on cleared land and local crafts could represent economic networks to be determined. The Charter should foster the creation of networks by encouraging artisans to federate, to plan the creation of labels guaranteeing the quality and origin of their products, and to carry impact assessments on the exploitation of the resources (Mission pour la création du parc de la Guyane 2006). In case the sale of goods produced increases, which then should be included in development objectives, it would be advisable to monitor the ecology of species through scientific research. This research would determine the sustainability of the resource (quantity and geographic distribution of the populations) and its sustainable exploitation (capacity for regeneration, picking technique) (Davy 2006).

How will these limitations to exploitation be determined, and how will the communities of inhabitants be involved in their determination? When the protection of plant or animal species is necessary to the subsistence of indigenous communities or to the maintenance of their traditional way of life, decisions concerning potential measures are taken by the Park Director. This decision-making capacity has been formalised in Article 4 of the Decree on the Creation of the National Park. The decision of the Director is, however, guided by the opinions of the Scientific Council and the Committee for Local Life. Therefore, for example, to what extent will the Wayãpi be prevented from hunting the collared peccary which is very symbolic to them (Grenand 1996)? Alternatively, will the Wayãpi be able to force the Park Board to take special measures to protect this mammal? One can expect heated debates in this regard.

Surprisingly, the Teko of Camopi do not feel very concerned for the time being by the management of natural resources. They expect that the main source of revenue will come from tourism which is authorised throughout the core area, as is the construction of light tourist infrastructure. In fact, the mayor (at the time) who had been thinking for a long time of building traditional houses or *carbets* for tourists, as well as the representative of the traditional authorities who was a professional boatman, expected much from the park. However, many questions have arisen. Until today, the Order of the *Préfet* issued in 1970 and revised in 1977, regulates access to the upper parts of the rivers in the *Grand Sud* ('Indian country'). This access is authorised by the *Préfet* and the Charter will need to establish whether this authorisation is still compulsory, whether the task of authorisation will fall to the Park Board, and whether the communities of inhabitants will be entitled to prevent tourists from entering their villages and hunting trails. Finally, the Charter will need to define whether the communities have first option to build tourist infrastructure and to regulate the potential associations between them and the tourism agencies based in Cayenne.

Sharing Decisional Jurisdiction and Reconfiguring Alliances

The preparation works for the creation of the park revealed tensions within Guyanese society, where actors often had conflicting expectations. In this light, the participative process has been particularly delicate.

The position of the Guyanese local officials has been ambiguous, to say the least. The conduct of these officials has always been ambiguous towards the metropolitan power, and towards local communities. This ambiguity, however, did not prevent pragmatic and once-off alliances from being created. During the consultations prior to the creation of the park, the fact that metropolitan France refused to confiscate Guyanese territory in favour of the Amerindians (and to the detriment of the Creoles), was often vigorously and even violently denounced: the park must be for all Guyanese people and its wealth must not only benefit the "micro-local resident populations". At the same time, the same elected officials were against Guyana building up stronger relations with neighbouring countries, Brazil in particular which is deemed too conquering, and with the Amazonian region represented by the ACTO[9]. More generally, many elected officials as well as representatives from the private sector fear that the park will impede the

---

9   This is in fact a view which is shared by the central state which systematically associates the *Préfet* with the external diplomatic initiatives of the president of the Regional Council.

development of Guyana, especially as regards gold mining which represents an important source of local income. Yet they acknowledge that the park's existence can lead to an improved structure and tenure of the land, to the development of infrastructures, increased profitability from ecotourism (or even ethno-tourism), and a more comprehensive strategy against illegal gold panning.

How were the grievances and concerns of the local governments and the local communities heard and potentially conciliated? According to the Act of 2006, the administration of a park is carried out by a Board of Directors which includes representation from local actors: the elected officials and the members chosen for their local expertise (owners, inhabitants, farmers, professionals, users and environmental conservation NGOs) hold at least half the seats, the other half being distributed between the state representatives and national experts (whether scientific or institutional). The nomination of the members of the Board of Directors and their numbers was determined on a case by case basis during the establishment of the park. Nonetheless, the law provides for *ipso jure* members: mayors of *communes* with more than 10% of their territory in the core area (this provision already existed in the Act of 1960), the présidents of the Departmental and Regional Councils, and the chairman of the Scientific Council of the Park.

For the National Park of Guyana, the situation is different in that parity between the state and the local governments was not adopted. The law holds that the mayors of the five *communes* concerned are *ipso jure* members of the Board of Directors. According to the Decree on the Creation of the National Park, the council consists of 44 members: 10 state representatives, 12 local government representatives, five representatives from the Amerindian and Maroon communities (Box 6.2) and 16 key players, plus one personnel representative. The diversity of the council members is certainly expected to promote a wide array of opinions. Local representation is in the majority compared to state representation, insofar as the local governments emerging from the decentralisation and the communities of inhabitants are included in this category. However, the local representation is far from being homogeneous, and alliances between the state and the various associations are changeable.

For example, certain local governments can oppose the central state without acting in the interests of populations living in the park; and it is likely that the state representatives are, on certain issues, more favourable to the interests of the communities than the representatives of the local governments. Moreover, the state can rely on the loyalty of the mayors by offering them various development perspectives (roads and other infrastructure). Generally, the institutional motivations and personal preferences of the members of the Board of Directors may not always coincide.

The place reserved for the communities in the decision-making structures is absolute, but their actual power is uncertain, particularly within the Board of Directors. Likewise, during the procedure for the adoption of the Decree on the Creation of the National Park, within the Steering Committee and during the public enquiry, the Amerindian and Maroon communities were consulted directly

Box 6.2   Reduced participation of the local communities

The Board of Directors only has five traditional authority representatives out of the 44 members making up the board. These representatives are provided for by Article 28 of the Decree. They are appointed by the *'grand man'* concerned or, failing this (and therefore when several ethnic groups are involved), by the meeting of the 'captains' and household heads of the territory, convened by the mayor of the *commune* concerned. They have been made official by the appointment Order (of the Minister of Ecology and Sustainable Development) of 1 March 2007: a representative from the traditional authorities of the village and hamlets of Papaïchton (Aluku); for Maripasoula, a representative from the traditional authority of the village (where a majority of Akulu live), and a representative from the traditional authorities of the hamlets of Upper Maroni (Wayana and Teko), which means one representative for two ethnic groups; for Camopi, a representative from the traditional authorities of the hamlets of the middle Oyapock, the hamlets situated on the banks of the Camopi River and the village (Wayãpi and Teko); and a representative from the traditional authorities of the hamlets of Upper Oyapock (Wayãpi).

and through representatives of the traditional authorities. However, their opinion was not actually enforceable by the French state. In addition, representatives from these communities will sit on the Committee for Local Life. Yet, whether legally or practically, consultations will need to be conducted but all opinions will remain purely consultative.

Despite these limitations, the fact that the Amerindian and Maroon authorities have been taken into consideration must be highlighted. Indeed, for a long time common law authorities (mayors of the *communes*) and the traditional authorities (those tolerated by the administration) have co-existed. Although 'captains' and *'grands mans'* are granted some governance functions, formerly by order of the *préfecture* and currently by order of the Departmental Council, their duties have not been clearly determined. Thus far their duties have covered land clearing, setting dates for traditional holidays and providing a policing structure, although this function has often been questioned. Disputes are submitted to arbitration before the traditional chiefs and when a decision needs be taken, as a rule, the mayor of the *commune* concerned must consult with the traditional leader. Despite the French government's refusal to introduce the notion of collective rights in French law, which would lead to the official recognition of communities interposed between the citizen and the state, certain customary laws of the local Guyanese communities are in fact implicitly recognised (Collective 1999).

Traditional leadership is recognised by the park, however, the formal management structures could entail a loss of authority, for the traditional authorities are somewhat underrepresented on the Board of Directors. Moreover, the interventions of the mayors (who can also be Amerindian or Maroon) and the representatives of the traditional authorities, based on their abilities and legitimacy, remains to be seen. The Decree on the Creation of the National Park does not provide for the legal recognition of customary law, which means that the tolerance that prevailed prior to the establishment of the park could be affected. And yet, in all the Amazonian states, the explicit integration of customary law into the management plans of protected areas constitutes a cornerstone of conservation policies (Filoche 2007a).

The special case of access to genetic resources, mentioned specifically in the Act of 2006, serves to illustrate the complexity of the situation involving the capabilities of all the actors concerned. It also serves to illustrate the tensions between metropolitan France and the Departmental Council of Guyane, as well as that between Creoles local officials and the local communities living in the park. This is an issue which affects the Guyana Amazonian Park in particular, since the other National Parks do not deal with this question.

While the French government did ratify the Convention on Biological Diversity, it did not implement Article 15 as regards accessing genetic resources and benefit sharing drawn from their exploitation. To the Brazilian and Bolivian governments, Articles 15 and 8j of the CBD mean that bioprospection activities must be subject to a benefit-sharing contract with the indigenous and local communities. This contract should be drawn up with their prior and informed consent. This applies as soon as bio-prospection concerns a genetic resource which has already been used as a communal biological resource, i.e. for which communities would have in one way or another contributed to its perpetuation, and have indicated a possible usage or location for it. Potentially, a contract could also be drawn up as soon as a resource grows on the lands occupied by these communities (Aubertin et al. 2007).

Currently, these provisions have not been implemented in France. Several cases of biopiracy have questioned the activities of French public research institutes. These institutes have been denounced by the Guyanese authorities, as examples of the plundering of Guyanese heritage by metropolitan France.

The procedure for accessing genetic resources and associated knowledge exists neither for Guyana nor metropolitan France. However, the Act of 2006 contains a surprising proposal for local officials to take over the functions conferred by the CBD upon the state. The regulations for accessing and utilising resources and for sharing the benefits will not be defined by legislation from France nor by a Park Board regulation, but will result from a proposal of the congress of elected officials from the Departmental and Regional Councils of Guyana, to be recorded in the Charter. Under Article L. 331–15–6, only the président of the Regional Council, after receiving the assent of the President of the Departmental Council, can issue authorisations to access the genetic resources of species sampled in the National Park, "without prejudice to the application of the provisions of the intellectual property code". While the orientations to be recorded in the Charter must expressly respect the principles of the CBD, "those asserted in Articles 8j and 15 in particular", we need to question the extent to which the Charter will take into consideration the communities and the expression of their prior consent, all the more since neither the law nor the decree mentions traditional knowledge[10].

Conclusion: An Ambiguous Design

The current shape of the park could be said to reflect the conflicts between all interested parties (See Plate 12). Indeed, although initially the Steering Committee had worked towards establishing a unique core area, now there are three. This fragmentation is presented as the result of late procedures of 'participative democracy' during the public enquiry, where these procedures and their results were much debated by scientists and NGOs. From an ecological point of view, this fragmentation does not take into account one of the fundamental general ecological laws that exponentially links the number of species to the surface area sheltering those species (Rosenweig 2007). Moreover, nothing indicates that the ZFAs will be connected in such a way as to enable the establishment of corridors between the three core areas[11]: it will depend on the Charter negotiations.

From a socioeconomic point of view, the layout of the park ignores the wish of the Wayana – which was probably expressed too late – to benefit from protection against the ravages of gold panning, by having their villages included in the core zones. These villages will therefore be part of the ZFA, provided the *communes* of Maripasoula and Papaïchon adhere to the Charter. This division is all the more worrying since it could be interpreted as granting *garimpeiros* easy access to gold washing sites. While gold washing is definitely forbidden in the core area, it can be authorised in the ZFA and may even be allowed upstream in rivers crossing one of the three core areas of the park, depending on what the Charter will enact. In this regard, the concept of clean and sustainable gold washing is far from reassuring (Collective 2005). Finally, deciding not to include in the core most of the areas bordering Surinam and the Parque nacional das montanhas do Tumucumaque in Brazil, opens up the possibility for uncontrolled transactions.

The Guyana Amazonian Park gave Guyanese officials an opportunity to assert their authority over a National Park and local communities. There is no doubt that local governments will indeed be controlling the drafting process of the Charter, since they will have a majority vote, and since their adherence to this text will ensure the proper functioning and sustainability of the park. Nevertheless,

---

10   Yet initially, thanks to an amendment introduced during parliamentary debates, it was acknowledged that the local communities had their own decision power and control over bioprospection through their traditional political authorities (Karpe 2007). This amendment did not hold: it was argued in particular that it was better if genetic resources were appropriated by the entire Guyanese community, and not simply by a few scattered communities.

11   See Carrière et al. and Bonnin, in this publication, about corridors and ecological networks.

they will still need to respect the Decree and the Act (as these are higher in the hierarchy of legal standards), although local governments will be sure to make the most of the leeway offered by these texts. While it is too soon to draw conclusions, the creation of the park was a missed opportunity to grant local communities legal status and unambiguous rights over their lands and resources. Having the entire implementation of the park's operation rely on a future negotiated Charter is a risky wager, in a context where local populations sometimes find it difficult to make their voices heard in relation to the state, to local governments and to economic or even ecological interests. Indeed, it is a risky bet concerning the benefits these populations should be drawing from the creation of the park, and concerning the conservation objectives that will be threatened by the economic imperatives of gold panning.

References

Aubertin C., Pinton F., Boisvert V. (eds.), 2007 – *Les marchés de la biodiversité*. Paris, IRD Éditions, p. 272.

Collective, 1999 – Les autochtones de l'outre-mer français. *Droit et Cultures*, Special issue, 37 (1), p. 316.

Collective, 2005 – *État des lieux de l'exploitation de l'or en Guyane*. Document from the Collective "Quel orpaillage pour la Guyane ?" p. 75.

Davy D., 2006 – *L'artisanat de vannerie dans les communes du sud guyanais. État des lieux ethnoécologique et socio-économique*. Report from IRD/Mission pour la création du parc de Guyane, p. 124.

Filoche G., 2007 a – *Ethnodéveloppement, développement durable et droit en Amazonie*. Brussels, Bruylant, p. 650.

Filoche G., 2007 b – La réforme des parcs nationaux français. Diversification des acteurs, redéfinition des compétences et des outils de gestion. *Revue Européenne de Droit de l'Environnement*, 3: 309–320.

Fleury M., Karpe P., 2006 – Le parc national de Guyane: un arbitrage difficile entre intérêts divergents. *Journal de la Société des Américanistes*, 92 (1–2): 303–325.

Giran J.-P., 2003 – *Les parcs nationaux: une référence pour la France, une chance pour ses territoires*. Paris, La Documentation française, coll. Rapports officiels.

Grenand P., 1996 – "Des fruits, des animaux et des hommes: stratégies de chasse et de pêche chez les Wayãpi d'Amazonie". *In* Hladik C.M. et al. (eds.), *L'alimentation en forêt tropicale. Interactions bioculturelles et perspectives de développement. Vol. II (Bases culturelles des choix alimentaires et stratégies de développement)*, Paris, Unesco: 671–684.

Grenand F., Grenand P., 2005 – Trente ans de luttes amérindiennes. *Ethnies*, 18 (31–32): 132–163.

Grenand F., Bahuchet S., Grenand P., 2006 – Environment and peoples in French Guiana: ambiguities in applying the laws of the French Republic. *International Social Science Journal*, 187: 49–58.

Karpe P., 2007 – L'illégalité du statut juridique français des savoirs traditionnels. *Revue Juridique de l'Environnement*, 2: 173–186.

Mission pour la création du parc de la Guyane, 2006 – *Parc amazonien de Guyane. Projet. Livret.* Cayenne, 49 p. + maps.

Pinton F., Aubertin C., 2005 – "Populations traditionnelles: enquêtes de frontière". *In* Albaladéjo C., Arnauld de Sartre X. (eds.), *L'Amazonie brésilienne et le développement durable. Expériences et enjeux en milieu rural.* Paris, L'Harmattan: 159–178.

Renoux F., Fleury M., Reinette Y., Grenand P., Grenand F., 2003 – L'agriculture itinérante sur brûlis dans les bassins du Maroni et de l'Oyapock: dynamique et adaptation aux contraintes spatiales. *Revue Forestière Française*, 55, Special issue: 236–259.

Rosenweig M. L., 2007 – La biodiversité en équation. *Les dossiers de La Recherche*, 28, août-octobre: 20–24.

Untermaier J., 2008 – Le parc amazonien de Guyane, huitième parc national français (décret n° 2007–266 du 27 février 2007). *Revue Juridique de l'Environnement*, 2: 135–155.

Chapter 7

From Amerindian Territorialities to "Indigenous Lands" in the Brazilian Amazon: The Yanomami and Kayapó Cases

Bruce Albert, Pascale de Robert, Anne-Élisabeth Laques
and François-Michel Le Tourneau

Protected areas, under 19 different statuses, cover almost 41% of the surface area of Brazil's Amazon region. As conservation areas, they are used by the state as a tool of land blocking which is supposed to prevent economic ventures, and therefore subsequent deforestation (Léna 2005)[1]. The inhabitants of these protected areas, when their presence is tolerated, are thus ascribed a stereotypical social immutability, as is often the case with so-called traditional societies. Yet, on the contrary, we could regard the capacity of these societies to constantly adjust their relationships to the natural environment and to social others, both locally and in a wider interethnic context, as enabling inhabited protected areas to play a significant role in the conservation of the environment. In this perspective, when the actors of social change manage collectively to control its dynamic, this can become a guarantee of environmental conservation.

To illustrate this point, we present in this chapter a study of two Amerindian groups from the Brazilian Amazon taking as examples the villages of Apiahiki and Moikarakô, respectively situated in the Yanomami and Kayapó indigenous lands. The territories of these two groups, traditionally unbounded, were recently marked out and legalised in the form of specific protected areas known as 'Indigenous Lands' (*Terras Indígenas*). On analysing the historical process which led to the official recognition of these areas, we were able to assess some aspects of the impact that such a transformation had on the local indigenous management of space and resources of the tropical forest. Through these examples, we try to highlight the way in which Amerindian societies invent forms of 'sustainable development' satisfying at the same time their own values and the exogenous demand for the conservation of their lands as protected areas.

1   The research behind this text was conducted within the framework of the UR 169 IRD-MNHN, of the trans-départemental incentive "Protected Areas" of the IRD and of the UR 169 partnership in Brazil with the Instituto Socioambiental of São Paulo (B. Albert) and the Federal University of Rio de Janeiro-Laget (P. de Robert, A.-E. Laques).

Amerindian Territories and Conservation in the Brazilian Amazon

Protected Areas with a Special Status

The legalised Amerindian territories of Brazil, called 'Indigenous Lands', benefit from a complex status of social, cultural and, indirectly, environmental protection. The legal framework of these protected areas is defined in Articles 20 and 231 of the Brazilian Constitution of 1988, which respectively allocate their ownership to the Federal Union and their exclusive usufruct to the Amerindian populations who occupy them.

The Brazilian Constitution defines 'Indigenous Lands' very broadly, encompassing areas occupied and exploited by Amerindian groups at a given time, as well as all other areas deemed necessary to their future physical and cultural requirements. This extensive definition has resulted in a significant increase in both the number and the surface area of Amerindian territories recognised in the Amazon region. Today, we can estimate their surface area to be 1,084,665 km², i.e. 21.7% of the so-called 'legal Amazon'[2] (See Figure 7.1).

Although their status does not explicitly relate to nature conservation, Amerindian territories have a fundamental significance in the Brazilian system of protected areas. Indeed, in Brazil the federal and state system of conservation units covers 1,000,020 km² in the Amazon, partially overlapping with several Indigenous Lands or reserved areas incompatible with environmental protection, such as military lands and gold panning reserves. In total, the surface area of these conservation units only represents another 20% of the 'legal Amazon' area, i.e. much less than that of the Indigenous Lands of the region if we consider the overlaps.

Recent quantification of what had already been perceived empirically by local actors has confirmed that the legal framework protecting these territories, and the presence of resident populations mobilised to protect them, act as major factors in the prevention of deforestation and forest fires (Nepstad et al. 2006). Furthermore, the creation of these areas also represents a cheaper prevention measure for the state[3]. In this light, it appears that the more effectively preserved areas in the Amazon, from the conservation point of view, are the territories occupied by the Amerindian groups and legally recognised as 'Indigenous Lands'. They should be thus considered as a fundamental form of environmental protection. In recent years, this conservation function of Indigenous Lands has certainly been

---

2   The 'legal Amazon' consists of 6 states of the north region of Brazil (Amapá, Pará, Roraima, Amazonas, Acre and Rondônia) as well as the new state of Tocantins, western Maranhão and northern Mato Grosso. This administrative region extends over around 5 million km² (almost 59% of the country's surface area).

3   Joint study of the American NGO The Nature Conservancy and the Co-ordination of Indigenous Organisations of the Brazilian Amazon (COIAB). See http://www.coiab.com.br/jornal.php?id=379.

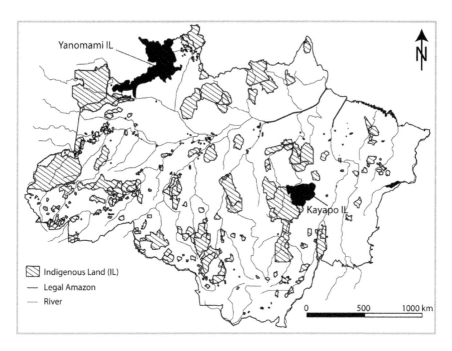

Figure 7.1   Indigenous Lands of the Brazilian Amazon

more valued by the Brazilian Ministry of Environment and the National Indian Foundation (FUNAI), as well as non-governmental actors. It has also been given more attention by Amerindian leaders who have integrated this argument into their demands for a better recognition of their land rights, giving them a broader access to the media and the national political scene (Albert 1993;1997, Turner 1999; Turner and Fajans-Turner 2006).

Protected Areas in the Face of Social Change

However, despite their considerable potential for nature conservation, many Indigenous Lands are subject to strong economic pressures that could put in jeopardy their integrity in the medium and long term. In these cases, environmental threats are both external and internal.

External threats correspond either to predatory, and mostly illegal, incursions aimed at exploiting the natural resources available in the Amerindian territories (precious wood, gold, diamonds, tin ore, etc.), to the construction of public infrastructure (roads, dams, electricity lines), or to the advance of the agricultural frontier (soya and cattle farming).

Internal threats relate to the consequences of the demographic and socioeconomic changes experienced by Amerindian societies that are increasingly coming into contact with the regional economic frontier. Population growth, habitat regrouping, increasing settlement and changes in the lifestyles and productive activities of these groups, can thus lead to the impoverishment of their knowledge about the forest, to a less diversified usage of their environment, and to the overexploitation of certain local natural resources. Finally, the risk of collusion between new Amerindian political elites and regional economic actors (such as illegal loggers or gold panners), with a view to opening up access to Indigenous Land resources, could aggravate the impact of external threats on these protected areas.

From this perspective, it is fundamental to discuss the political and socioeconomic conditions under which the Amerindian territories will be able to sustain their role as protected areas in the Brazilian Amazon. Although in other studies, we examined this problem from a regional perspective (Albert 2001; 2004; Le Tourneau 2006), we will here deal with the issue at the local level. As such, we will try to account for the way in which the internationally well-known Yanomami and Kayapó Amerindian societies rearrange their territorial space and the use of its natural resources at village level according to external threats, social changes, and new sustainable development opportunities.

The Yanomami and Kayapó: From Amerindian Territory to 'Indigenous Land'

Yanomami Indigenous Land

The presence of the Yanomami in the Parima Mountains, on the border between Brazil and Venezuela, has been progressively revealed to the Western world by explorers and boundary commissions since the 19th century. However, this region being particularly difficult to access, neither Brazilian nor Venezuelan society had attempted to enter and exploit the area before the 1970s.

In the 1950s, the establishment of Catholic and Protestant missions created the first permanent points of contact between the Yanomami and the Whites, on the outskirts of their territory. However, interactions remained very limited until two major waves of expansion of the Brazilian national and regional economic frontier affected the Yanomami territory and society.

The first chapter of this expansion took place between 1973 and 1976, with the construction of a section of the Perimetral Norte Highway, which was intended to run parallel to the Trans-Amazonian Highway on the left bank of the Amazon River. The construction project was launched with no regard for the Yanomami, and led to the decimation or disappearance of several villages on the eastern outskirts of their territory. It also led to the spreading of epidemics, particularly measles and influenza, on a vast scale, even reaching very remote villages. Owing to a lack of financial resources, the military government of Brazil of the time finally abandoned this highway construction project, averting a massive invasion of the Yanomami territory.

However, this massive invasion did occur 10 years later after it was revealed that the alluvial deposits of the rivers irrigating the Yanomami lands from the Parima Mountains were rich in gold. 40,000 gold panners invaded the centre of the indigenous territory between 1987 and 1990, giving origin to the largest gold rush of the 20th century. Again, the consequences were catastrophic for the Amerindians, with many villages decimated by the spread of malaria imported by the gold panners and disseminated from the numerous placers in the area, accompanied by complete disorganisation of the traditional system of natural resource usage. It was estimated at the time that 15% of the Yanomami population disappeared during those three years.

This tragedy, strongly denounced by national and international NGOs as well as by the press, led to the intervention of the Brazilian Attorney General which, in turn, resulted in the creation of a protected territory for the Yanomami. Thus, in 1992, a vast area of around 96,500 km², called the 'Yanomami Indigenous Land' (See Plate 13), was officially demarcated for the Yanomami and Ye'kuana ethnic groups.

Even if this decision did not solve all the problems, it established a legal framework for state relevant administrations to take action to deal with the low-key but still recurrent invasions of gold panners, and to contain the deadly spread of malaria among the Yanomami. Thus, despite still precarious health assistance, the situation improved for most Yanomami communities. Today, they represent about 250 local groups with a total population of around 17,000 people, which is now clearly expanding.

Beyond their continuous fight against territorial invasions and for the improvement of healthcare, the Yanomami are now increasingly concerned with the management of their territory and its natural resources. In this regard, they recently created (2004) a political association in Brazil to represent their communities, called the Hutukara Yanomami Association. Today this association, in partnership with a São Paulo-based environmental NGO called Instituto Socioambiental (ISA), seeks to define its own sustainable social development orientations, and to implement these thanks to the national and international financing of educational 'projects' (e.g. network of schools teaching in the Yanomami language), social projects (e.g. training of association executives, local radio network and cultural disclosure) and environmental projects (reforestation and economic alternatives).

However, the integrity of the Yanomami territory is still confronted with a number of serious external threats. These are due to the persistent incursions of gold panners around the Parima Mountains, whose camps and placers, among other ills, propagate infectious and parasitic diseases (Albert and Le Tourneau 2005). These health issues regularly endanger entire communities, forcing them to spend relatively long sedentary periods close to health posts, thereby paralysing their productive activities and the cycle of their agricultural work in particular.

Furthermore, since 1978, projects for agricultural colonisation and cattle ranching implemented in the western part of the state of Roraima by federal and, thereafter, regional land institutes, opened up an encroaching frontier on the eastern border of the Yanomami territory (enlarged by an accompanying movement of illegal land occupation). Although this frontier is yet not particularly dynamic, it has reached the border of the Indigenous Land (Le Tourneau 2003; Albert and Le Tourneau 2004) and, in some cases, settlers and farmers have begun to cross it (e.g. in the south-eastern region of the Ajarani River). In addition to their predatory use of resources belonging to the Yanomami (hunting, fishing and logging), these settlers and farmers systematically practice an extensive slash-and-burn agriculture. In doing so, dry seasons have been getting more intense in this region each year, causing giant forest fires – as was the case in 1998, 2003 and 2007 – thereby affecting directly, and irreversibly, the biodiversity of this area (Barbosa 2003).

Finally, 54% of the surface area of the Yanomami Indigenous Land is the subject of 640 applications for industrial prospecting or exploitation permits, submitted to the Brazilian National Department of Mineral Production by various public, private, national and multinational companies (Ricardo and Rolla 2005). These applications and recently proposed bills aimed at legalising regulated mining activities in Amerindian territories in Brazil constitute, in the medium and long term, a considerable challenge for the protection of the natural environment of the Yanomami territory.

Kayapó Indigenous Land

The first information about the Kayapó, who call themselves *Mebêngôkre*, dates from the 19th century. At that time they formed three major groups hostile to one another, living in a region where the plateau savannas meet the forests of the plains between the Araguaia, Tocantins and Xingú Rivers, south of their current location. After enduring conflicts, first with slave traffickers and later with rubber tappers and Brazil nut gatherers, the Kayapó refused any peaceful contact, even with the other ethnic groups of the region, and progressively migrated towards less accessible forested areas to the north and the west.

The first Kayapó who decided to engage in less conflictual relationships with the Whites were rapidly decimated by epidemics. Most of today's Kayapó are the descendants of groups who only accepted peaceful contact with the other populations of the region during the 1950s, after a long period of resistance. These first interactions, accepted and experienced differently by each village, were in most cases promoted by the regional authorities, to satisfy the repeated demands from local colonists eager to exploit, unhindered, the lands and resources of the region (e.g. feline skins, Brazil nuts and gold). Generally, the dynamics of internal conflicts, scissions, migration movements and wars (exacerbated by the recent acquisition of firearms) were particularly intense for all the Kayapó during the first half of the 20th century (Turner 1998).

During the 1960s the pacification of internal relations, progressive access to medical care and the beginning of a population recovery, did not lessen the threat of the advance of local populations towards Kayapó lands. The opening of the Brasília-Belém Highway in the 1970s and the Xinguara-São Felix do Xingú Highway in the 1980s, located respectively on the eastern and northern margins of Kayapó territory, brutally intensified contact with the regional economic frontier, leading to the fragmentation of traditional territory and, in some regions, facilitating the mass invasion of illegal gold prospectors.

During the 1980s and 1990s the Kayapó fought to protect their lands mainly by expelling these gold diggers and mobilising against dam projects on the Xingú River. These battles were led by notorious Kayapó leaders and were supported by many environmental and indigenous NGOs, as well as some show business personalities. These actions caught the attention of the national and international media, and accelerated the legal recognition of the different Kayapó territories in the form of 'Indigenous Lands' (Turner 1999). This is also the period during which certain Kayapó villages embarked on partnerships for socio-environmental projects supported by NGOs or private companies. The current Kayapó population is estimated at 7,400 individuals occupying around 20 villages with relative political autonomy, and spread across seven Indigenous Lands, each with different ecological characteristics. At the time, most of these Indigenous Lands had already been officialised[4]. The Kayapó territory, situated on the two banks of the Xingú River, a southern tributary of the Amazon River, represents 130,000 km² in total (See Plate 14). Nevertheless the legal recognition of the Kayapó's territorial rights did not stop the threats to their lands.

During the 1990s the participation of the Kayapó in the illegal exploitation of mahogany wood (*Swietenia macrophylla*) raised many controversies among the sympathisers and defenders of Amerindian and Amazonian causes – people who had always recognised the 'ecologist' reputation of this ethnic group – and among *Mebêngôkre* society. Indeed, internally, which types of relationships to adopt with the Whites were never unanimously agreed upon, provoking various revolts by the 'common people' opposed to the initiatives of certain leaders dealing with illegal loggers (Fisher 2000), and provoking scissions or manoeuvres to marginalise villages overly involved in timber trade. The proceeds from the sale of mahogany wood were sometimes used to finance Indigenous Land monitoring operations, as well as for initiatives of political communication with the outside, whilst unsold wood was simply stolen by the loggers. At the time, to engage in a campaign alongside ecologists while also negotiating timber deals was not necessarily perceived as contradictory. All transactions with the illegal loggers have been stopped for several years now, due as much to new alliances uniting the majority of Kayapó villages with two major NGOs (Instituto Raoni and Conservation International) (Schwartzman and Zimmerman 2005), as to the fact that mahogany resources have been progressively depleted.

The first decade of the new millennium has been characterised by an abandonment of all links with the predatory activities of loggers and gold diggers, and by a

---

4   The "Ratification" (*homologação*) is the last stage of the process for the legal recognition of an Amerindian territory in Brazil.

multiplication of Amerindian associations, founded in almost every village, with the purpose of obtaining public and private funds for the implementation of economic, social and cultural projects. Today these sustainable development projects play a central role in the internal politics of the *Mebêngôkre* and, complemented by the income generated from a few old age pensions[5] and salaries (of teachers, nurses and FUNAI employees), enable the Kayapó, like the Yanomami, to buy basic consumer goods, although on a much larger scale. These projects reinforce relationships between ethnic groups and the regional NGOs which have been promoting dialogue between villages, within a framework of common initiatives (Zimmerman et al. 2006). They are elaborated around three themes: health and education, sustainable economic alternatives and territorial monitoring.

Today, Kayapó lands are under pressure from the agricultural frontier, with annual fires threatening their eastern and northern borders, and with the establishment of illegal pastures in the forest. There is thus a risk that these threats will compromise living sites, disrupting the traditional utilisation of the environment, and that they will be intensified by the advance of the regional economic frontier with the opening or rehabilitation of main highways such as the BR163[6].

The Forest Space in Apiahiki (Yanomami) and Moikarakô (Kayapó)

After presenting the creation process of the Yanomami and Kayapó Indigenous Lands, we will outline, in this section, the pattern of natural resources in two villages of these 'inhabited protected areas'. As a matter of fact, whether in the medium or the long term, the protection of the natural environment of Indigenous Lands will depend on these local models and their capacity to adapt to new situations.

For over 50 years, in order to keep an acceptable distance from the regional economic frontier, the Yanomami and Kayapó villagers have developed complex migration dynamics which represents one of the keys to their adaptation to historical changes. In this light, we can consider that the internal reorganisation of their models of space utilisation, after the recent official recognition of their lands as protected areas, plays the same role as far as adapting to new environmental issues is concerned.

---

5   Amerindians, like other members of the Brazilian rural population, have access to a 'rural retirement pension' called *aposentadoria rural*.

6   The inhabitants of the Indigenous Land of Bau, a Kayapó territory situated more to the west, already had to suffer the consequences of this intensifying territorial pressure when, in 2003, after 10 years of conflicts with illegal loggers, gold panners, cattle ranchers and politicians from the region, and after a long court case, their territory was reduced by the Brazilian government by nearly 300,000 ha (Inglez de Souza 2006).

Routes, Places and Resources: Traditional Use of the Forest Space

The configurations we studied in the villages of Apiahiki and Moikarakô, which result from the adaptation of traditional systems to present circumstances, are very similar in many instance. While in this section we describe some of their characteristics, we will examine the extent to which space and mobility can constitute fundamental variables for the sustainability of these Amerindian natural resource exploitation systems.

The forest and gardens provide both villages with their basic requirements. Food resources (through gathering, fishing and hunting), materials for the construction of houses and the manufacture of various implements (e.g. bows, baskets, tools and canoes), ritual objects (e.g. ornaments and clubs) as well as certain medicinal and other plants with active substances (psychotropics, stimulants and fishing poisons), all come from the forest. Food plants (manioc, banana, sweet potato, yam, sugar cane, corn and pawpaw, among others), tobacco, blowpipes, cotton (for hammock and ornaments in Apiahiki) and fibres (for rope making), as well as plants used as remedies or for propitiatory purposes or for witchcraft[7], are all extracted from gardens. The forest and gardens thus constitute the hub of the economic and social life of each village.

The gardens are usually located close to the village, although the availability of good soils can sometimes encourage villagers to cultivate much farther afield (as much as two km in the case of Moikarakô). A network of paths links all the gardens to the village. This network often crosses former agricultural sites left fallow but still used to collect remaining productive plants for subsequent transplantation. It also runs through the forest to specific sites containing high densities of fruit trees, or to areas for hunting large game, catching small animals and fishing with piscicidal plants (See Plate 15).

In the case of the Yanomami villages, such as Apiahiki, this network of forest paths exits the village, crosses the gardens and connects a complex set of hunting, fishing and collecting sites. In its maximal extension this web of trails ends at various camp sites where long-term collective hunting and gathering expeditions are organised (Albert and Le Tourneau 2007). In the case of Moikarakô, the current network of paths used regularly by the younger, more sedentary generations is less extensive than that of the Yanomami. However, the sites occupied by the current villagers, and the paths used by them during the last decades, have been extended widely within the Kayapó Indigenous Land. In time, many old paths are likely to be reactivated by the Kayapó, as has happened already within the framework of a border monitoring and Brazil nut marketing project.

The two villages make use of their rivers differently. Today, the Kayapó of Moikarakô make intensive use of the river alongside the village, fishing upstream as well as downstream over six km or so with their traditional canoes, and farther

---

7   Concerning the use of plant resources among the Yanomami and the Kayapó, see Albert and Milliken (2009) and Posey (2002) respectively.

afield when fuel is available for their aluminium motorboats. This was not the case for their first village, situated near a minor river where women and children were more diligent in fishing, and men went hunting more often. Moreover, the river constitutes an important means of communication with other villages situated several days upstream or downstream. The land paths are rarely used to visit other distant communities, as the villagers prefer to go there by plane or boat.

In Apiahikɨ, the presence of a larger stream has not profoundly changed the attitude of the Yanomami towards the river. Indeed, the Yanomami population come from the highlands of the Orinoco-Amazon interfluve and are not traditionally familiar with navigation techniques. Although line fishing, which the Yanomami perceive as a masculine activity, contributes significantly to the food input of the village, it is not considered as prestigious as hunting, and travelling on the river (with dug-out canoes or aluminium boats belonging to the health services) remains a limited activity. As a result, the inter-village visits and the bulk of the economic activities are carried out via the complex network of paths criss-crossing the region.

In Moikarakô as well as in Apiahikɨ, the extensive networks of forest paths and rivers enable the villagers to easily access the resources they need to survive, maintaining a low pressure on the available natural resources, and protecting the ecological dynamics of exploited areas. These networks exclude vast areas from the reach of humans, which can then serve, for example, as reproduction and refuge areas for game. Furthermore, the two communities do not make intensive use of the forest space; neither seeks to identify and exploit the available resources exhaustively, nor to mark out their territory and systematically appropriate it. The Kayapó and the Yanomami only take what they need from the forest at specific times, and do not seek to generate surpluses for storing or marketing purposes.

It is crucial to understand that in the Kayapó or Yanomami systems, the forest is not seen as a separate entity in drastic opposition to the village or garden. On the contrary, all three spaces make up a whole into which the way of life of humans (and non-humans alike) is smoothly integrated. By not recognising wild and domestic or nature and culture as separate realities, the Kayapó and the Yanomami leave no room for a narrative focused on nature protection: the forest is experienced as a component of a global cosmology and a primary condition of human existence. The destruction of the environment is simply unthinkable, except in the context of a cosmological disruption and the disappearance of humankind. Of course, the absence of ontological distinction between nature and society among the Kayapó and the Yanomami does not prevent them from making political compromises with the ecological conceptions of their non-indigenous allies proposing sustainable development projects, even when such conceptions are culturally incompatible with their own[8].

---

8   See Albert (1993) for an ecological translation of Yanomami shamanic conceptions on the devastation of the forest by the gold panners at the end of the 1980s, and Albert (1997: 193–198) on Kayapó ritual eco-ethnicity.

Indigenous Lands and Restructuring of Traditional Spatial Systems

The intensification of the economic contacts with the Whites and the official delimitation of the ethnic territories have put new pressures on the traditional model of land use in Apiahiki and Moikarakô, thereby challenging the capacity of that model to adapt to a new historical context. However, the reticular and sporadic exploitation of the forest resources by these communities has proved to be particularly flexible insofar as, keeping unused spaces in reserve, it can always offer alternative strategies of use and therefore fit the ecological, social and political contexts as needed. The redeployment of the path networks in these available spaces is thus adjusted depending on the need and according to a varying combination of settlement and mobility at any time.

In the case of Apiahiki, the possible movements of the community are limited by two kinds of restrictions, external and internal. They are first limited in the south by the presence of a health post of the National Health Foundation (FUNASA), on which their members depend for healthcare and access to certain essential goods (such as metal tools, cooking pots, clothing, salt, etc.). They are also restricted on different sides by the path networks and hunting camps of several neighbouring communities. In this context, being confronted with the increased scarcity of their resources, the villagers of Apiahiki decided first to redeploy their path networks in a long northern 'corridor' (thereby reusing a former migration route), and second to establish a temporary collective house (Sinatha 2) at the end of this corridor. This new round-house, surrounded by gardens and inhabited during several months of the year, constitutes a 'second home' also used as a base for collective hunting and gathering expeditions, in an area rich in game and fruit trees. This 'bi-cephalous' two-house residential set up constitutes in fact a reduplication of the basic traditional residential Yanomami layout (which consists of one collective habitation and one or several satellite forest camps). This dual variant enables the villagers to optimise the sustainable use of the available resources in a delineated area, by playing on the spatiality and temporality of productive activities. As such, the flexibility of the Yanomami model of land use leads to the skilled and sustainable management of the region's forests, without having to resort to the conceptual and institutional framework of our conservation units that are culturally exogenous and socially constraining (See Plate 16).

The village of Moikarakô has only been established on its current site for the past five years. Its territory is situated in the centre of the Indigenous Land, far from the borders of the protected area and, as such, is little exposed to external pressures. The villagers still don't feel the impact of their sedentary lifestyle on the availability of resources around the village. Moreover, the recent opening of new gardens on the opposite bank, and the ritual collective hunting sessions in the territory of the neighbouring village, show that the boundaries separating the territories of these communities remain flexible and negotiable inside the Indigenous Land, which was not always the case in the 1990s during the episode of illegal mahogany exploitation. Finally, the original site of Moikarakô is still used as a second village. This new configuration of land use constitutes also a kind of transposition of the traditional Kayapó model of circulation between the main village and several smaller satellite villages and forest camps (Verswijver 1992): a model which today has become obsolete. In the same light, we must also take into consideration the 'hosting facilities' (e.g. associations, missions, hospital, etc.) where people of Moikarakô temporarily stay in neighbouring urban centres which, like the forest, have become a source of supply. Thanks to all these changes, the Kayapó are able to overcome the limitations imposed by the creation of the protected area and keep the extensive territoriality valued by their way of life (de Robert 2004).

Indigenous Lands and Development Actors

Besides the Amerindians, a number of actors exercise a certain influence on the sustainable management of the natural environment of the Indigenous Lands, and on the conservation of their resources. These actors, whose interventions (or projects) are linked to the economy of the local communities, are of three types: government administrations, national or foreign NGOs and indigenous associations. They can intervene in these protected areas collaboratively, independently or competitively. The relative weight of each actor tends to fluctuate according to the times. Nonetheless, we find a marked tendency towards the expansion of the NGO sector, and a certain withdrawal from the (direct) action of the public sector. Moreover, Amerindian associations, which have been multiplying since the 1990s, have seen their influence among NGOs grow constantly (Albert 2001).

Governmental Organisations

Guaranteeing the integrity of Indigenous Lands depends upon the National Indian Foundation (FUNAI), the Brazilian indigenous administration which in turn depends on the Department of Justice[9]. The territorial control exercised by FUNAI *via* its network of 'indigenous posts' remains fairly theoretical, simply because it does not have the budget to implement an effective monitoring system likely to secure the borders of the country's 672 Indigenous Lands (1,115,236 km^2).

Despite its obvious institutional limitations FUNAI often remains an important actor locally, and is sometimes the only state referent when faced with private regional economic interests (which it has the greatest difficulties to control when it is not their accomplice), with missionary abuses and with the initiatives of NGOs sometimes more inspired by corporatism than a concern for Amerindian interests.

---

9   Since the "Indian Status" of 1973 the Amerindians in Brazil, considered as minor, are under the guardianship of FUNAI. This FUNAI anachronistic attribution should be revoked when a new piece of legislation, currently discussed by the Brazilian Congress (the "Status of Indigenous Societies"), will be adopted.

Moreover, since the 1970s it is often on the basis of the first territorial surveys of the FUNAI, whether or not these are disputed and/or endorsed by Amerindian political mobilisations and indigenous NGOs, that the legalisation of many Indigenous Lands has been implemented. Finally, many Amerindian groups – like the Kayapó and, to a lesser extent, the Yanomami – have succeeded in appointing some of their own people as heads of FUNAI posts, thereby appropriating a function emanating from the state administrative structure to increase their own autonomy, as well as using it as a platform for their political agenda.

At the level of resource management, the historical initiatives of FUNAI often turned out to be disastrous, whether these were based on the establishment of agricultural colonies (focused on the 'indigenous post' institution) completely unsuitable for the Amerindian communities (e.g. the cultivation of rice and black beans and cattle farming), or on forest product extraction projects based on the traditional Amazonian model of paternalistic exploitation. A few remnants of these initiatives are still found in certain regions of the Yanomami territory (agriculture and cattle breeding on the lower portion of the Mucajaí River, and harvesting of *Leopoldina piassaba* palm fibre in the Rio Negro region). However, since the 1980s the network of 'indigenous posts' of FUNAI in the Yanomami Indigenous Land has decreased to such a point that today its impact there is very limited.

The Kayapó, on the other hand, found themselves involved in many 'community development projects' organised by the local administration of FUNAI established in the nearby small town of Redenção, which for a long time was the sole interlocutor in the context of Kayapó outside relations (Inglez de Souza 2006). For the main part, these projects supported collective expeditions to gather Brazil nuts which were then marketed regionally, with FUNAI covering, for example, the fuel costs for the transport of the nuts by boat. The projects also led to the opening of 'community gardens' intended for subsistence farming, with FUNAI supplying seeds (rice and beans) and tools. Somehow, this last initiative has had a long-lasting impact on Kayapó agriculture. Indeed, rice has today its own place in the organisation of cultivated spaces and in the Kayapó botanical nomenclature, and has become a high valued food (which is why it is often bought in town). However, in Moikarakô rice does not replace local crops in any way, and has not given rise to any ambition to intensive farming or marketing. Its cultivation remains confined to the framework and periodicity of 'rice projects' financed by FUNAI, at least when they are carried out. On the other hand, machetes and wheelbarrows obtained through the project find other uses, to cultivate sweet potatoes and transport stones for the traditional ovens, for example. As such, the 'community development projects' of FUNAI are generally twisted by traditional strategies aimed at acquiring industrial goods and taken over by the logic of local politics. Nevertheless their impact is weakening, insofar as the means of indigenous administration are themselves decreasing.

NGOs and Indigenous Associations

NGOs involved directly in the two Amerindian territories can be distinguished according to two categories: national and international. The Pro-Yanomami Commission (CCPY) is an example of a national structure although its environmental intervention is still modest, working in the Yanomami Indigenous Land. Founded in 1978, this Brazilian NGO has a strong historical legitimacy through its involvement in the fight for the legalisation of the Yanomami Indigenous Land, its support for the political organisation of the Yanomami and the establishment of many field projects over the past three decades, particularly concerning education and health. At the environmental level, the initiatives of CCPY are more recent (1990s) and intentionally kept in a low-key mode.

The strategy of this NGO was to restrict its interventions to a few regions of the Yanomami territory that were likely to undergo environmental degradation in accordance with the increasingly sedentary lifestyle and the demographic expansion of certain communities, or to regions already degraded by external interventions. In this context, the Apiahiki community is involved in a fruit tree planting project (local and imported trees) and an apiculture project, both aimed at reinforcing the availability of food resources to the community around its main collective house. These small projects are conceived more in a spirit of complementarity between this principal habitation and its satellite (Sinatha 2) than as an intervention aimed at steering the local economy as regards sustainable development. They are financed mainly through funds from the Brazilian Department of Environmental Affairs through its programme of environmental demonstration projects (PDA)[10] and, from 2000, through it sub-programme of indigenous demonstration projects (PDI).

In the Kayapó Indigenous Land the international NGO Conservation International (CI) has been present since 1992, where it initially financed a research project called the Pinkaiti Project which only concerned the village of A'Ukre (Schwartzman and Zimmerman 2005). The objective was to create a kind of sanctuary within the Indigenous Land, i.e. an area of 80,000 ha situated about 15 km from the Kayapó village, and, in exchange for taxes and salaries paid to the Indians, reserved exclusively for conservation and biological research.

The stated or even sole objective of CI remains the protection of the environment. However, in its interventions with the Kayapó this organisation seems to be moving towards a more long-term investment in favour of sustainable development, taking into account the priorities imposed by *Mebêngôkre* society. At the end of long negotiations and an expansion of its themes and places of intervention since 2000, CI today finances the projects of the two most important Kayapó associations: the Associação Floresta Protegida, which was created with the support of CI and to date includes most of the villages of the Kayapó

---

10   This programme was created in 1995 with the financial support of the international co-operation of the PPG7 countries (Pilot programme for the protection of tropical forests) and the German international co-operation in particular.

Indigenous Land; and the Instituto Raoni which looks after the interests of the Kayapó villages of Mato Grosso.

These two associations have recently begun work on a common project called the 'Kayapó Project'. The main objectives of this project are to monitor the territory of the Indigenous Land (in collaboration with FUNAI), implement alternative economic projects and, in the long term, promote education and health. Paradoxically it seems that these initiatives are likely to be beneficial from the conservation point of view primarily because they are already appropriated by the Kayapó, whose motivations are far from 'saving the environment'. Indeed, the meetings with all the *Mebêngôkre* leaders intended to support their action within the project, the collective training sessions and the resumption of traditional long-distance expeditions for territorial monitoring or to seek economic alternatives, seem to favour a movement of political revitalisation and land re-appropriation. In fact, a major success of the Kayapó Project would be to skilfully serve the internal politics of the Kayapó so as to federate around a single objective the representatives of all the Kayapó communities scattered throughout various Indigenous Lands.

The Kayapó villages are used to enjoying extensive political autonomy and have for some years been involved in creating many associations with often competing interests. Most of these associations were founded from 2000 onwards, within the context of the decentralisation of health services for Amerindians promulgated by the Brazilian government National Health Foundation (FUNASA) in 1999. They don't develop projects directly linked to the protection of the environment and their health initiatives (e.g. construction of health posts, wells and water adduction systems, and pest control operations) seem to result in a more sedentary lifestyle in the Kayapó Indigenous Land. However, for the Kayapó themselves these initiatives constitute a vehicle for the expansion of their social space towards urban centres, and a means to solve internal conflicts. The political disagreements between communities, and even within villages, generally give rise to the creation of new associations in a process that somehow supplants the scissions and migrations of the past (de Robert 2010). This dynamic opens new social spaces outside the Kayapó Indigenous Land: whereas in 1998 there was only one association in Redenção, today there are a dozen, with their head offices situated in the small regional towns of Redenção, Ourilandia, Tucumã, Colider, Marabá and São Felix do Xingú (Inglez de Souza 2006).

Unlike the Kayapó, the Yanomami have deliberately chosen to establish a single political association, Hutukara, which represents all the regions of the Yanomami Indigenous Land (November 2004). Its head office is situated in Boa Vista, the capital of the state of Roraima. Choosing such a structure (i.e. one association with 27 regional delegates) is very much the result of a long struggle, from 1977 to 1992, against the fragmentation of the territory, at first by CCPY and since the 1980s by the Yanomami themselves. Since its foundation, Hutukara has been focusing on its institutional and political consolidation (establishment of head office, establishment of radio network in the regions, administrative and legal training of managerial staff, operations of disclosure and political intervention). Although it does not currently have any direct environmental management activity in the Yanomami Indigenous Land, its objective is to take over CCPY projects in the medium term. However, the bulk of its activities on the ground concerns territorial surveillance, its staff constantly checking, by radio, for invasions and environmental degradation in the 27 regions of the Yanomami Indigenous Land. Therefore, as in the case of the Kayapó, the fact that the Yanomami adopted an association shows their will to protect their territory and conserve their tropical forest environment threatened by the regional economic frontier. It also shows their determination to arm themselves politically to support this land struggle by expanding their traditional social and political space towards neighbouring urban centres.

Conclusion

The Indigenous Lands of the Brazilian Amazon were not initially created to protect the environment of the region but to guarantee Amerindians their historical rights over protected areas in which they could maintain their social structures and control potential changes (Brazilian Constitution of 1988). But, despite the lack of environmental concern of these constitutional provisions, the fact remains that due to the low population density of Amerindian groups and the low impact of their productive systems, the Indigenous Lands of the Amazon today function like protection islands in the face of the economic frontier encroaching on the region. On the other hand, it is obvious that the sustainability of this function of conservation can only be guaranteed if the model of Amerindian use of natural resource is not subject to radical transformations, to the point that their characteristically low ecological impact is called into question.

In this context, we can wonder about the ecologically perverse effects of some ongoing social changes in Amerindian territories, such as the population nucleation and the settlement of villages, or the increasing economic contacts of some of their inhabitants with the regional frontier. We know also that these changes could be reinforced by the official recognition of these areas as Indigenous Lands and the exogenous policies and interventions associated to this process. On the other hand, we observed that the Amerindian communities we studied, for whom the forest environment constitutes a vital element, are involved in a constant process of readjusting their models of land use so as to counterbalance the pernicious effects of endogenous social changes and externally induced pressures. We have mentioned the diversity and complexity of these adjustments in Apiahikɨ and Moikarakô, where several strategies have been developed for that purpose, such as the redeployment of forest path networks, the establishment of systems of double residence, the cultural twisting of governmental and non-governmental development projects, the expansion of association networks and the appropriation of new inter-ethnic social spaces.

These adjustments testify to the remarkable adaptability and creativity of the social and economic Amerindian systems. They also testify to the strength and continuity of the fundamental parameters of space and time underpinning the organisation and reproduction of the 'sustainability' of these societies. In this light, dispersion and mobility appear as the crucial constants of a variable geometry guaranteeing their sustained use of the tropical forest space and its resources. Considering the geographic and ecological importance of Indigenous Lands in the environmental protection system of the Brazilian Amazon, it is crucial that the sustainable development policies designed for the region fully understand and take into account these multiple dimensions of Amerindian territoriality.

References

Albert B., 1993 – L'or cannibale et la chute du ciel. Une critique chamanique de l'économie politique de la nature. *L'Homme,* 126–128: 353–382.

Albert B., 1997 – Territorialité, ethnopolitique et développement. À propos du mouvement indien en Amazonie brésilienne. *Cahiers des Amériques Latines,* 23: 177–210.

Albert B., 2001 – Associations amérindiennes et développement durable en Amazonie brésilienne. *Recherches Amérindiennes au Québec,* 31 (3): 49–58.

Albert B., 2004 – Les Indiens et l'État au Brésil. *Problèmes d'Amérique Latine,* 52: 63–84.

Albert B., Le Tourneau F.-M., 2004 – "Florestas Nacionais na Terra Indígena Yanomami – Um cavalo de Tróia ambiental ?" *In* Ricardo F. (ed.), *Terras Indígenas & Unidades de Conservação da natureza.* São Paulo, Instituto Socioambiental: 372–383.

Albert B., Le Tourneau F.-M., 2005 – Homoxi : ruée vers l'or chez les Indiens Yanomami du haut Mucajaí (Brésil). *Autrepart,* 34: 3–28.

Albert B., Le Tourneau F.-M., 2007 – Ethnogeography and resources use among the Yanomami Indians: towards a 'reticular space' model. *Current Anthropology,* 48 (4): 584–592.

Albert B., Milliken W., 2009 – *Urihi a. A terra-floresta Yanomami.* São Paulo, Instituto Socioambiental, p. 207.

Barbosa R.I., 2003 – "Incêndios florestais em Roraima: implicações ecológicas e lições para o desenvolvimento sustentado". *In* Albert B. (ed.), *Fronteira agropecuária e Terra Indígena Yanomami em Roraima,* Documentos Yanomami 3, Brasília, Comissão Pró-Yanomami: 43–54.

Fisher W., 2000 – *Rain Forest exchanges. Industry and community on an Amazonian frontier.* Washington, Smithsonian Institution Press, p. 222

Geffray C., 1995 – *Chroniques de la servitude en Amazonie brésilienne. Essai sur l'exploitation paternaliste.* Paris, Karthala, p. 185.

Inglez de Souza C., 2006 – "Kayapó. As relações com a sociedade envolvente". *In* Ricardo B., Ricardo F. (eds.), *Povos Indígenas no Brasil 2001/2005.* São Paulo, Instituto Socioambiental: 501–505.

Jerozolimski A., 2006 – *Caracterização do uso e disponibilidade de Castanha-do-Pará (Bertholletia excelsa) no território da comunidade Kayapó de Kikretum, sul do Pará*. Fieldwork report, Conservation International-Brasil, p. 15.

Léna P., 2005 – "Préface". *In* Albaladejo C., Arnauld de Sartre X. (eds.), *L'Amazonie brésilienne et le développement durable. Expériences et enjeux en milieu rural.* Paris, L'Harmattan, coll. Amérique latine: 7–16.

Léna P., Geffray C., Araujo R., 1996 – *L'oppression paternaliste au Brésil. Lusotopie*, Special issue, 19: 103–353.

Le Tourneau F.-M., 2003 – "Colonização agrícola e áreas protegidas no oeste de Roraima". *In* Albert B. (ed.), *Fronteira agro-pecuária e Terra Indígena Yanomami em Roraima.* Documentos Yanomami 3. Brasília, Comissão Pró-Yanomami: 11–42.

Le Tourneau F.-M, 2006 – Enjeux et conflits autour des territoires amérindiens d'Amazonie brésilienne. Problèmes d'Amérique Latine, 60: 71–94.

Nesptad D., Schwartzman S., Bamberger B., Santilli M., Ray D., Schlesinger P., Lefebvre P., Alencar A., Prinz E., Fiske G., Rolla A., 2006 – Inhibition of Amazon deforestation and fire by parks and Indigenous Lands. *Conservation Biology,* (20) 1: 65–73.

Posey D. A., 2002 – *Kayapó Etnoecology and Culture.* London, Routledge, p. 283.

Ricardo F. (ed.), 2004 – *Terras indígenas e unidades de conservação da natureza. O desafio das superposições.* São Paulo, Instituto Socioambiental, p. 687.

Ricardo F., Rolla A., 2005 – *mineração em terras indígenas na amazônia brasileira.* São Paulo, Instituto Socioambiental, p. 179.

de Robert P., 2004 – Terre coupée. Recompositions des territorialités indigènes dans une réserve d'Amazonie. *Ethnologie Française,* 34 (1): 79–88.

de Robert P., forthcoming – "Conflitos, alianças e recomposições territoriais em projetos de desenvolvimento sustentável: experiências da Terra Indígena Kayapó (Sul do Pará)". *In* Araújo R., Léna P. (eds.), *Desenvolvimento sustentável e sociedades na Amazônia..* Belém, MPEG.

de Robert P., Laques A.-E., 2003 – "La carte de notre terre". Enjeux cartographiques vus par les indiens Kayapó (Amazonie brésilienne). *Mappemonde,* 69: 1–6.

Schwartzman S., Zimmerman B., 2005 – Alianças de conservação com povos indígenas da Amazônia. *Megadiversidade,* 1 (1): 165–173.

Turner T., 1998 – "Os Mebêngôkre-Kayapó. História e mudança social, de comunidades autônomas para a coexistência interétnica". *In* Carneiro da Cunha M. (ed.), *Historia dos Índios do Brasil.* São Paulo, Cia das Letras: 311–338.

Turner T., 1999 – La lutte pour les ressources de la forêt en Amazonie : le cas des Indiens Kayapó au Brésil. *Ethnies Documents,* Special issue, 13 (24–25): 115–148.

Turner T., Fajans-Turner V., 2006 – Political innovation and inter-ethnic alliance. Kayapó resistance to the developmentalist state. *Anthropology Today,* 22 (5): 3–10.

Verswijver G., 1992 – "Entre village et forêt". *In* Verswijver G. (ed.), *Kayapó, Amazonie. Plumes et peintures corporelles*. Tervuren, Musée royal de l'Afrique centrale: 11–26.

Zimmerman B., Jerozolimski A., Zeidemann V., 2006 – "Kayapó apostam em alternativas econômicas sustentáveis". *In* Ricardo B., Ricardo F. (eds.), *Povos Indígenas no Brasil 2001/2005*. São Paulo, Instituto Socioambiental: 506–508.

Chapter 8

Pastoralism and Protected Areas in West and East Africa

Jean Boutrais

The relationship between protected areas and pastoralism is becoming a major preoccupation for environmental administrators in West, Sahelian and Sudanese Africa, although this concern is relatively new to them, whereas in East Africa it is already well established and raises important economic and political issues. Certain authors (Bourn and Blench 1999: 2) go as far as estimating that, in West Africa, the numbers of large fauna are so diminished that the issue of their co-existence with cattle is no longer a problem. In East Africa, they identify the area of competition between fauna and livestock as a long strip (Figure 8.1) that stretches from Sudan through Kenya and all the way to Botswana, Namibia and South Africa. According to this map, there is no such competition in West Africa, where the landscape is dominated by pastoralists and their herds.

Although this perception is generally valid, it requires further consideration. Indeed, the issue of pastoralism concerns a whole set of protected areas in the Sudan region and a few Sahelian reserves, from eastern Senegal to northern Cameroon and Chad. It is true that in West Africa, the protection of the environment was not a priority during the colonial period. As for Sahelian pastoralism, it benefited from the constant support of the colonial administration, whether from a veterinary perspective or in terms of the hydraulic equipment used in the pastures. In this domain, there used to be a consensus between pastoralists and the colonial administration.

On the other hand, during the same period in East Africa, the colonial discourse was already challenging local populations who were held responsible for environmental destruction. According to a belief that was widely held among British settlers, there were simply too many people and too many cattle. This notion led to initiatives involving the forced sale of cattle and the limitation of cattle stocking on pastures, making the colonial administration unpopular with many pastoralists.

In West Africa, it was only at the end of the 20th century, during the great droughts, that pastoralism was accused of degrading the vegetation and the soils, causing the desertification of the Sahel. While large protected areas were created in the savannah to serve as shelters for fauna under attack from farmers, pastoralism is now currently portrayed as the main threat to conservation. At the time, the narratives of conservation officials resolutely adopted anti-pastoral positions. Although they are more recent than their East African counterparts, have West African environmental policies not become more radical in relation to pastoralism over time?

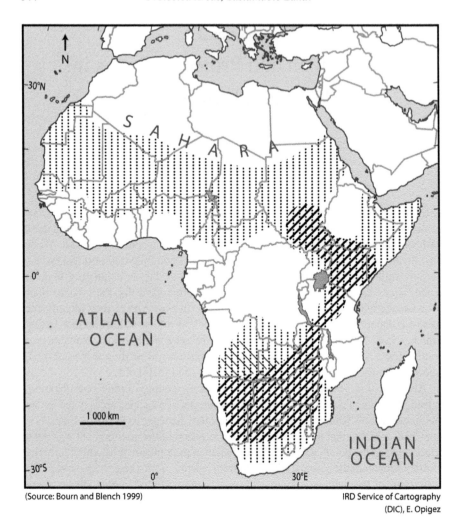

(Source: Bourn and Blench 1999)

IRD Service of Cartography
(DIC), E. Opigez

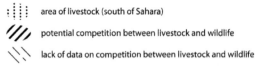

area of livestock (south of Sahara)

potential competition between livestock and wildlife

lack of data on competition between livestock and wildlife

Figure 8.1   Livestock and wildlife in Africa

The viability of nature conservation through the use of protected areas is only possible, in the long term, if it is recognised and accepted by local actors (Blench 2004: 10). This text reflects the problems of the relations between pastoralists and protected areas in Sub-Saharan Africa. It is not a study on protected areas per se, but on their relationships with a group of local actors. Even if we take important ecological processes into account, our text is focused on the geographical dynamics on the outskirts of protected areas in particular. In this perspective on conservation and pastoralism, our approach is based on pastoralism.

Geographic Comparison between Protected Areas and Pastoral Areas

The historical geography of pastoralism over a century or so, offers an initial explanation of the difference between West and East Africa as to the relations between pastoralists and protected areas.

Spatial Gap or Combination of Pastoralism and Conservation

At the beginning of the 20th century, a map of livestock raising areas in French Western Africa drawn up by a veterinarian illustrates that these areas mainly covered the Sahelian zone (Pierre 1906). In French Sudan at the time, the southern boundary went around the Macina, up to the outskirts of Ouagadougou, then to Fada and Say. There was no large livestock raising (i.e. Zebu herding) in Côte d'Ivoire, nor in what is today south-west Burkina Faso. 40 years later, the geography of the cattle breeds, and therefore of large stock raising, had not yet changed significantly (Doutressoulle 1947: map 6). Whereas the north boundary ended up being pushed further into Mauritania and Niger, the south boundary remained stable during the first half of the century. On the whole, and except for a few deviations in either direction, this boundary corresponded to that of the distribution of tsetse flies, which prevent the prolonged presence of zebu in the wooded savannah.

The large protected areas created in the 1920s and 1930s (i.e. the W National Parks and the National Parks of Bénoué and Faro) and, later, at the end of the 1940s and 1950s (e.g. the National Parks of Bouba Njida and Niokolo-Koba), were at the time situated outside pastoral areas. Most of these forest and wildlife reserves were part of areas devoid of populations, areas corresponding to no man's lands between chieftainships and pre-colonial kingdoms. These intermediary zones were the locus of raids that often forced the populations to scatter even further. Yet, these 'open' areas were exploited by mobile groups who made the most of the abundant natural resources (Benoît 1988). Not only did these small groups share the same refusal to be subjected to a vast and centralised power, but they also shared the same ideology of equal access to resources, and a respect for nature. More prosaically, for the local administration, the creation of reserves in these in-between areas offered a solution to the difficulty of controlling and 'subjugating' them due to their small populations.

On the other hand, the no man's lands between controlled areas were often sought after by pastoralists, particularly the Fulani. Indeed, these lands were often covered with abundant pastures, and enabled pastoralists to evade taxation and cattle requisitioning by certain chiefs. Consequently, these bush areas, in the ecological as well as the political sense, often served as migration corridors for pastoralists, e.g. from Macina to Sokoto.

However, these migration flows and the emergence of Fulani pastoralism at the end of the 19th century and the beginning of the 20th century often occurred outside of areas that were later converted into reserves. In this regard, M. Benoît (1999) was able to recreate the stages of the establishment of the Fulani people in this sector of the Niger valley, with reference to the W National Park. According to the author, the Fulani herds were, at the time, situated about 100 km from the park when it was created in 1926. Pastoralists did not venture into the savannah of the current park, even for the seasonal moving of livestock, whereas hunters and even farmers tried to settle there, especially after the droughts of the beginning of the 20th century. The areas selected by the colonial administration for the implementation of reserves to protect the environment, areas that had been the seat of wars before the colonial era, remained dangerous. Whereas in these areas farmers feared the dangers linked to isolation, pastoralists dreaded these areas for their detrimental effects on cattle, due to tsetse fly infestations. Either way, cattle were small in number and their ownership remained precarious, even among the Fulani.

On the other hand, in East Africa, it is acknowledged that pastoralists, particularly the Maasai, have been coexisting with wildlife for a very long time, i.e. more specifically since the emergence of pastoralism towards 4000 BP, according to Bourn and Blench (1999). In the 1890s, this part of the continent experienced a catastrophic outbreak of rinderpest followed by a smallpox epidemic, and at the same time became affected by colonial conquests. While the former ruined Maasai pastoralism, the latter ended their political expansion. Although the consequences of human and animal depopulation on the natural environment have been the subject of debates (Ford 1971; Waller 1988), it is acknowledged that this led to the spread of scrub vegetation in the savannah, where the dominant herbaceous vegetation was previously maintained through fire and grazing. Shrub invasions in turn prepared the way for tsetse flies, which transmit bovine trypanosomosis. Despite the more of less rapid reconstitution of their herds, pastoralists did not re-occupy all their old pastures and remained confined to healthier zones (Homewood and Rodgers 1991). Vast areas that had become unhealthy and were inhabited by wildlife were then established as reserves from the beginning of the 20th century, particularly in the south of Kenya (the Maasai-Mara Reserve) and in the north of Tanzania (which later became the National Park of Serengeti). But the establishment of these reserves meant the alienation of lands and sometimes the expulsion of herds (as in the Serengeti) at the expense of the Maasai pastoralists.

Whereas the rinderpest of the 1890s also wrought devastation on the livestock of West Africa, losses were rapidly restored and the pastoral areas were not subjected to a high abandonment rate such as in East Africa (Boutrais 2007a). Because protected areas were remote and disconnected from pastoral areas, large wildlife did not come into contact with the herds of the pastoralists. Nevertheless, this isolation ceased during the latter decades of the 20th century.

Pastoralism vs Protected Areas in West Africa: Recent Developments

In West Africa, the end of the 20th century was marked by the regular expansion of pastoral areas in the savannah of the south, over the entire Sudanese and Sudano-Guinean zones, from Senegal to North Cameroon and Central Africa (Figure 8.2).

Studies conducted in this regard were able to scientifically analyse and understand this expansion in Côte d'Ivoire for example (Bernardet 1999). In 1985, although the Fulani pastoralists only occupied an area adjacent to the border of Burkina Faso, several years later, their grazing lands cover almost the entire north of the country, thereby entirely surrounding the National Park of Comoé.

The expansion of pastoralism towards the south was caused by ecological as well as political factors. In some countries, the colonial administrations forbade pastoralists to settle in the south for veterinary reasons and for fear of conflict between the populations. Post-independence the new administrations exercised less control over pastoral migrations. Moreover, these were facilitated by veterinary treatments that became increasingly effective at overcoming the tsetse fly problem. In any case, the droughts of the 1970s and the 1980s actually served to free, in a natural way, the savannah from tsetse flies. Whereas the droughts impoverished and even destroyed Sahelian pastures, they improved the pastoral quality of the savannah and even opened them up to zebu herding.

The great droughts led the pastoralists to leave their traditional Sahelian pastures and move towards the savannah of the south (Bassett and Turner 2007). These acted as an ecological refuge for populations who were completely destabilised and, as a result, began to enter protected areas. The reconstruction of the pastoral pressure scenario in the sector of the W National Park in Niger revealed the determinant role of the 1973 and 1984 droughts. Before 1973, Nigerian pastoralists moved their herds for rainy season grazing towards the north. In 1973 and 1984, in this exceptional context, many pastoralists resorted to a panic migration towards the south and the W National Parks. After that, the movements towards the south became regular and came to be perceived as 'normal'. Some sectors of the W National Parks became included in the pastoral space of the Fulani pastoralists according to the phenomenon of habituation and adaptation (Boutrais 2007b).

Climate crises were not the only reasons for which pastoralists were forced to change their spatial practices. The agricultural invasion of the pastures, a slower, more insidious and irreversible phenomenon, ended up greatly affecting pastoralism. Over time, pastoral systems became challenged, as in Western Niger: the seasonal movements of the herds towards the valley of the Niger River in an orientation perpendicular to the valley became impossible, and were replaced by meridian movements towards the W National Parks (Amadou and Boutrais 2005).

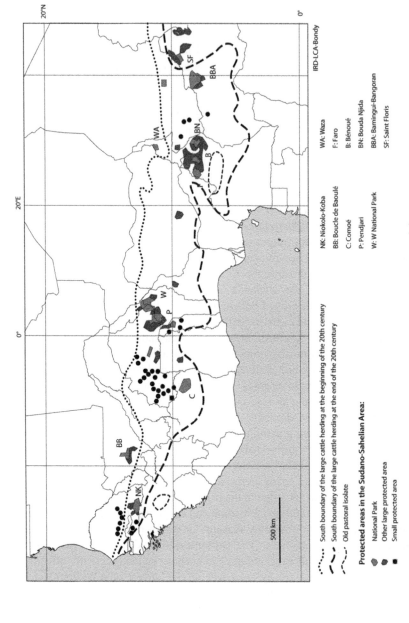

Figure 8.2 Protected areas and large cattle herding in West Africa

········· South boundary of the large cattle herding at the beginning of the 20th century

‒ ‒ ‒ South boundary of the large cattle herding at the end of the 20th century

·‒··‒ Old pastoral isolate

Protected areas in the Sudano-Sahelian Area:

National Park

Other large protected area

Small protected area

NK: Niokolo-Koba
BB: Boucle de Baoulé
C: Comoé
P: Pendjari
W: W National Park

WA: Waza
F: Faro
B: Bénoué
BN: Bouda Njida
BBA: Bamingui-Bangoran
SF: Saint Floris

IRD-LCA-Bondy

Generally, protected areas became attractive to pastoralists, not only due to their pastures but also because of their lack of cultivated areas.

With the deployment of pastoralism in the savannahs, pastoralists ended up surrounding most protected areas in the Sudanese zone. In several countries, transhumance movements reached the peri-forest savannah of the Guinean area, as was the case in Benin (Houndagba et al. 2007: 332). As a result, the issue of coexistence of cattle and wildlife, which had been marginal in West Africa during the middle of the previous century, became a problem. In this regard, the relations between pastoralism and wildlife can be assimilated with those experienced for a long time in East Africa. But there is a difference concerning the main conceptions of relationships between pastoralism and protected areas. Indeed, whereas in East Africa the focus is on the relations between wildlife and domestic cattle, in West Africa, the debates particularly concern the consequences of grazing for the protected vegetation (Fournier and Millogo-Rasolodimby 2007: 38).

Cattle In and Around Protected Areas

By introducing their herds into Sudanese protected areas, pastoralists intend to make the most of the more diverse and denser herbaceous cover than that found in non-protected areas. In protected areas, the intervention of foresters is carried out essentially in favour of the trees and wildlife, and is not overly concerned with protecting the grass (Kiéma 2007: 210). Consequently, for many herders, a cattle grazing on herbaceous vegetation in protected areas is compatible with the protection of ligneous vegetation. On the contrary, in the Sudanese area, grazed grass reduces the aggressiveness of bush fires for small shrubs.

In short, pastoralists enter protected areas not only in response to real pastoral pressures, but also because they believe grazing does not harm the protected areas concerned (Kiéma and Fournier 2007: 448).

Pastoral Intrusions into Protected Areas

The way cattle enter protected areas differs according to their geographic location in relation to livestock raising centres, and also according to the health context, particularly the degree of glossina infestation. Rangers' reactions against herders in protected areas vary, partly according to the periods of illegal presence.

For instance, the transhumant herders in northern Cameroon only enter the edges of the large parks of Bénoué during the dry season. Although they try to stay there until the beginning of the rainy season, the increase in glossina pressure forces them to leave the area rapidly.

Conversely, the timetable of occupation for three small protected areas in western Burkina Faso, shows that their presence is almost permanent during the year for two of these protected areas (Kiéma 2007: 175). Two periods mark maximum presence: at the end of the dry season and the beginning of the rainy season (May–June), and during the transition period between the rainy and dry seasons (October). Whereas the first period corresponds to a critical phase of cattle feeding, the second enables herders to prevent their herds from damaging crops just before, and during, the harvest. In brief, pastoralists living next to these small protected areas cannot do without them, not only because they provide fodder, but also because they offer a way of avoiding conflicts with farmers. They also offer a form of pastoral security, by relying on the more relaxed surveillance of the rangers, compared to that of the farmers. As for the return of transhumant herders, they enter the forest reserves sometimes with the intention of moving through them (Figure 8.3), particularly at the beginning of the rainy season. In this light, the small protected areas of western Burkina Faso act, in turn, as refuges for neighbouring herds, and as host areas for transhumant herders (Kiéma and Fournier 2007: 450).

In south-western Chad, the Yamba-Berté Forest Reserve, like the small protected areas of Burkina Faso, offers pastoral security. Indeed, on their arrival in the region during the 1970s, the first Fulani pastoralists had enough pastures at their disposal between the reserved lands. In time, with the increase in rural populations and the expansion of cultivation to produce more cotton, the pastures simply became less available. In this context, the Yamba-Berté Forest Reserve represented a decisive advantage: the agro-pastoralists established on the outskirts of the forest herded their cattle there permanently during the rainy season, to prevent them from damaging the cotton fields. Today, the forest reserve acts as a pastoral refuge. Again, during the dry season, local agro-pastoralists send their cattle to graze there, but only for the day. In short, several groups of pastoralists affected by the advance of agriculture have been falling back on the Yamba-Berté Forest Reserve. Ultimately, we could say that it is the presence of this protected area which enables the pastoralists to survive (Sougnabé et al. 2004).

Throughout West Africa we appreciate the complexity of the motivations and strategies of pastoralists vis-à-vis protected areas. In Benin, the herders who come from the Sahel of Niger are not the only ones who enter the W National Park. Pastoralists settled in the vicinity of the park also send their herds there, so as to avoid conflicts with their neighbours who cultivate cotton. Whereas the Sahelian enter the park during the dry season the local pastoralists go there during the rainy season (Toutain et al. 2004). Pastoral incursions into protected areas often remain seasonal. As they take place mostly during the dry season, when resources are in short supply, they create even more competition between domestic cattle and wildlife than exists during the rainy season. In this regard, pastoral incursions are even more strongly punished by the Forestry Department that, moreover, spots cattle more easily during the dry season.

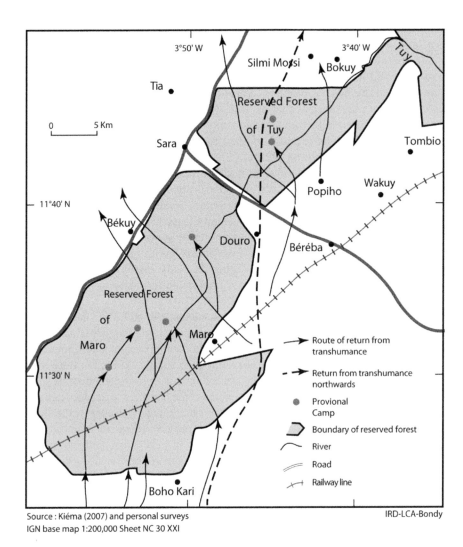

Source : Kiéma (2007) and personal surveys
IGN base map 1:200,000 Sheet NC 30 XXI

IRD-LCA-Bondy

Figure 8.3   Return   from   transhumance   through   reserved   forests (Burkina Faso)

Ecological Dynamics on the Outskirts of Protected Areas

Pastoralists are drawn to the areas surrounding protected areas for several reasons. If the need arises, they can easily make quick incursions into protected areas and withdraw from them just as fast. Cultivation there remains relatively limited due to the low-density population and to the fact that crops can be destroyed by wildlife. The zones of contact between protected areas and 'open' areas experience complex ecological processes linked to discontinuities between increasingly differentiated environments. Certain edge effects turn out to be favourable to pastoralism in the short term, while others are responsible for ecological dangers specific to these areas.

On the outskirts, pastoralists make the most of the plant species flowing from protected areas (seen here as reservoirs). Whereas intense grazing tends to impoverish the flora of pastures, protected areas result in the outskirts being repopulated by plant species that became rare or had disappeared from remote pastoral areas. Where protected areas constitute a source of grass seeds, the edge effect created on their entire periphery attracts pastoralists[1].

However, the prolonged stay of herds on the outskirts of protected areas exposes them to health risks. Indeed, large protected areas in the savannah are centres of tsetse flies and represent a threat to cattle. Although wildlife is not affected by bovine trypanosomosis, zebu herds are highly sensitive to it. Admittedly, the current incidence of this ecological constraint on cattle is difficult to evaluate, especially since the aridification of the environment has been mitigating glossina infestation in the savannah. The use of insecticides has also given more freedom to pastoral movements. Nonetheless, during the rainy season, glossina density increases to such a point that it makes it impossible for herds to stay in large protected areas.

For a long time entomologists have known that the zones on the edge of protected areas represent health risks for cattle. Recently, trapping glossinas along the gallery forests in the south-west of Burkina Faso confirmed their concentration on the edge, between forests and agro-pastoral areas (Bouyer 2006: 58; Bouyer et al. 2006). In the intermediary zones, the tsetse flies benefit from reproduction sites as well as a high number of hosts. Moreover, these flies manifest a trophic learning capacity, jumping from wildlife hosts to domestic cattle. The edges of gallery forests, just like the outskirts of protected areas, are also fragmented landscapes characterised by high natural gradients (whether in vegetation, temperature or hygrometry). In fragmented landscapes, glossinas are concentrated on the edge of dense vegetation. In homogenous landscapes, they scatter randomly, which mitigates the risk of domestic cattle being stung (Bouyer 2006: 120). In short, the creation of protected areas contributes to developing concentrations of glossinas which then constitute a threat to cattle.

------

1   In his research on protected areas of western Burkina Faso, S. Kiéma (2007) verified the hypothesis according to which a plant biodiversity gradient is a ratio of the distance to a protected area. Concerning ligneous vegetation, contrary to this hypothesis, Kiéma found that their density on recent fallows is lower when in proximity to protected areas (from 1 to 4 km) than when farther away (from 7 to 11 km). Kiéma accounts for this contradiction through the long term effects of intense grazing that, in the Sudanese savannah, provokes the well-known effect of scrub spreading. On the other hand, the herbaceous layer does indeed increase when close to protected areas. Young fallows close to protected areas benefit from colonisation by perennial herbaceous species and the *Andropogon gayanus* species in particular, which is highly appreciated by cattle.

The health issue in general is a subject of debate between conservationists and pastoralists. The first accuse domestic cattle of transmitting diseases to wildlife in situations where they coexist. Thus, in 1984, the buffaloes of the W National Parks supposedly died in great numbers of the rinderpest, a disease which was introduced by herds of zebus. Similarly, distemper, which is highly contagious, supposedly affected carnivores after being spread by the dogs of cattle farmers (Toutain et al. 2004). Yet, Fulani pastoralists keep almost no dogs, unlike the villagers (and the Tuareg). Conversely, the transhumant herders say that, in addition to trypanosomosis, buffaloes transmitted foot-and-mouth disease to their cattle. It is true that pastoralists do not really fear this disease whereas, for the international veterinary authorities, it represents a decisive element of discrimination.

In order to forbid the incursions of herds into areas reserved for wildlife, various measures restrict access to the outskirts established as buffer zones. Ideally according to the conservationists, the buffer zone surrounds the protected area in order to minimise contact between the transformed area and that reserved for nature. The aim of the buffer zone is, primarily, to prevent agricultural activities from spreading up to the boundaries of the protected area. Pastoral activities are also restricted there. The many hunting areas around the national parks of northern Cameroon and the forest reserves extending the Park of Niokolo-Koba in Senegal, illustrate the same buffer zone logic, without actually being officially called buffer zones.

The buffer zone *per se* is implemented when the outskirts of a protected area are established according to a specific layout. Thus, the Tamou Total Faunal Reserve in the north of the W National Park in Niger can only be used for grazing by the herds of cattle belonging to pastoralists already living there. Similarly, the aim of the project for land management situated on the edge of the Comoé National park in Côte d'Ivoire, was to establish a buffer zone between the Park and the customary users, including herders (Bassett 2002) who interpret these access restrictions as *de facto* extensions of the protected areas, at the expense of the pastures. The issue between conservation and pastoralism tends to be displaced from the protected areas per se to their outskirts. Even when no official buffer zones have been created, the Forestry Department seems to extend its control over cattle herds beyond the protected areas, with a view to preventing illegal entries.

> *In Niger, the foresters have created a "reserve" outside the W National Park. It was to prevent farmers from cultivating there. The herds could go there, up to the boundaries of the park. But now, the foresters no longer want the herds to enter the reserve. They fine huge amounts of money: 200,000 Francs, 500,000, 1 million ... What's the purpose of the reserve?* (Interview in Kollo, Niger, October 2005).

The exclusion of pastoralism from protected areas results in the creation of confrontation zones. An alternative to this is to have wildlife coexist with the cattle. As a result, the health effects of the edge situations are diluted in space. The cohabitation of wildlife and cattle defines what some authors already call "third generation" protected areas (Sournia 1998). In this scenario, non-fragmented areas replace wildlife sanctuaries; in other words, exceptional environments dedicated only to wildlife give way to ordinary environments, making room for cattle and wildlife.

Conservation and Cattle-Wildlife Coexistence

Nature conservationists often assert that the coexistence of cattle and large fauna causes a reduction in fauna numbers. They explain that competition is created between cattle and the wild herbivores, where both exploit the same ecological niches and manifest the same grazing behaviours. In this instance, zebus are in direct competition with wild herbivores. Another debate which constitutes one of the main controversies, and disrupted the implementation of the first protected areas in East Africa, concerns that fact that domestic cattle contaminate wildlife with contagious diseases, and vice-versa (McKenzie 1988). In fact, the only reliable and quantified data concerning these coexistence issues come from this region, yet they are contradictory.

Generally, the experts say that the issues created by Maasai pastoralism affecting large wild herbivores such as zebra, wildebeest and gazelle, are in fact harmless. In Kenya, the grid of their seasonal movements is identical (Western 1994). Based on the finding that the ecology of pastoralism and that of wildlife are interwoven and historically compatible, development plans integrating the conservation of wildlife and the socioeconomic development of the pastoralists have been implemented, as was the case for the wetlands of Amboseli (Western 1982). An increase in the numbers of elephants in this sector during the 1970s and the 1980s (while they were collapsing in the surrounding areas) has been attributed to the fact that the Maasai took an interest in the benefits of tourism. On the contrary, a nature park with 'strict' boundaries would not have protected the fauna as efficiently because, first, this type of park could not have retained all the fauna inside its boundaries throughout the year; and second, because it is acknowledged that the exclusion of pastoralists from reserves dedicated only to fauna does not contribute to its protection (Western 1994).

Studies conducted in Tanzania in the pastoral area of Ngorongoro, led to different results concerning animal populations in situations of coexistence (Arhem 1985; Homewood and Rodgers 1984; 1991). In a sector where the bovine livestock was to share the same resources with wildlife, both groups ended up sharing separate parts of the same space: the cattle herds occupied the highlands of the conservation area while wildlife was mostly concentrated in the plains (Arhem 1985: 55). In 1960 the number of wildebeests, large antelopes similar in size, needs and ecological strategies to bovines, crumbled following an epizootic of rinderpest. Subsequently it increased sharply until 1980, and remained stable thereafter. The authors surmise that this recovery was due to the vaccination of cattle against rinderpest, which in turn helped curb the propagation of the disease, although they attribute this recovery in the number of wildebeests mainly to the fact that cattle were excluded from their pastures.

Ultimately, the number of cattle declined on the whole. For the authors, this decline resulted from the herds' intense physical stress from years of grazing on low quality pastures infested with ticks. Moreover, the pastoralists of Ngorongoro could no longer move their herds during the rainy season, towards neighbouring plains occupied by huge herds of wildebeests. These antelopes are host of the malignant catarrhal fever virus, a highly contagious enzooty at that time of the year, and fatal for cattle. As a result, pastoralists avoided moving their herds onto the same pastures as those of the wildebeests, even though they represent a source of high quality fodder. Generally, the wildebeests are suspected to be host of tick-related diseases, these being increasingly feared when cattle herds remain sedentary during the rainy season (Homewood and Rodgers 1984). The pastoralists, who are aware that wildlife can represent a serious epidemiological risk, make sure to keep their cattle well away. In the Ngorongoro area, wildlife and livestock coexist on a wide scale, but do not mix locally (Homewood and Rodgers 1984).

In the general context of wildlife scarcity in West Africa, a few local situations have been favourable to progress when wild animals coexist with pastoralism. In Niger, the giraffe reserve of Kouré ensures the specific regeneration of giraffe numbers in an agro-pastoral environment. Here, a sharing of the forage resources occurs between cattle (i.e. the grazers), and the giraffes that browse on tree leaves. The distribution of salt by the pastoralists to cattle attracts giraffes that try to take advantage of the situation. The prolongation of the coexistence results in the semi-domestication of that species (Luxereau 2004).

The pastoralists who co-exist with protected wildlife can suffer negative consequences for their livestock. Yet, researchers who thought about the future of Ngorongoro deem that such coexistence is still preferable to exclusion, for pastoralists as well as for wildlife (Homewood and Rodgers 1984). Here, a few interventions in favour of pastoralism are intended for the time being, but a compromise aimed at the joint occupation of the area is recommended in the long term. This is an option which has never been envisaged in West Africa.

For wildlife, a beneficial consequence of the presence of cattle concerns predators. In the W National Parks for example, lions, that have become more numerous, can still fairly easily take wild prey during the dry season, at the edges of scarce watering holes. On the contrary, during the rainy season, wild herbivores scatter thanks to the multiplication of small watering holes. Therefore, during this period, it is relatively easier for the lions to occasionally prey upon cattle herds. They do this by coming out of the W National Parks and sometimes even by crossing the Niger River. Naturally, pastoralists have strong opinions regarding this.

Conversely, recent estimations show that in East Africa there has been a net reduction in wildlife numbers, particularly in Kenya, during the last two decades of the previous century (oral information from Homewood, 2004). However, this reduction is not due to pastoralist pressure but to the development of mechanised agriculture on a large scale. This form of agriculture, unlike pastoralism, destroyed the vegetation cover indispensable to wildlife. In eastern Burkina Faso, the current expansion of cotton cultivation, which is considerable, is also leading to agriculture spreading towards protected areas. Similarly, in northern Benin, cotton fields extend right up to the limits of the W National Park. In the coming years, this agricultural pressure will further transform the vegetation cover, unlike pastoralism.

Exclusion or Cattle-Wildlife Coexistence Policies

In West Africa, the Fulani pastoralists often claim that their cattle can coexist with the large wild herbivores (van Santen 2008: 283). They would even accept the occasional loss of cattle to predators, and agree to only seek to eliminate such predators were they to consider their losses excessive. The Fulani pastoralists are not particularly good hunters, nor do they consume bush meat, except for a few specialised lineages. In East Africa, the Maasai pastoralists also tolerate the coexistence of their cattle with wildlife. For example they do not prevent wildlife from approaching water points or salted places by erecting thorny enclosures. Camps are established at a distance from such places, so that cattle and wildlife can access these key resources in turn. This management of the area leads to a successful wildlife-cattle relationship (Homewood and Rodgers 1991: 192). In the past, for the Maasai of Kenya, wildlife was like a 'second cattle'. They used to rely on it when a drought had wiped out their cattle (Western 1994). They explained that they "used to milk" the wildlife when they no longer had any cows.

Are pastoralists' viewpoints taken into account and enacted by planning policies? In West Africa, all the Departments of Water Affairs and Forestry forbid domestic herds to graze inside protected areas. In fact, up until the 1990s, this policy was relatively un-enforced, for lack of means of surveillance and due to the frequent isolation of protected areas (Figure 8.3).

> *In the past, I used to move my cattle into the Maro Forest Reserve where I spent the entire dry season. Now, it's forbidden. So I move my cattle much farther, in the Lobi area where there is still some bush. I don't understand the grazing prohibition in the forest reserve. Grazing does nothing to the vegetation. To come back from the Lobi area, there are cattle tracks. But towards Boho, there are many fields. I walk along the Maro Forest Reserve, between the forest and the fields. From Douro, I cross the Tuy Forest to end up at Silmi Mossi.*
> (Survey in Tawremba-Bondoukuy, Burkina Faso, December 1995).

During the severe Sahelian droughts of the 1970s and the 1980s, pastoralists from Niger found refuge in the W National Parks, and were not immediately chased away by the rangers (Figure 8.4).

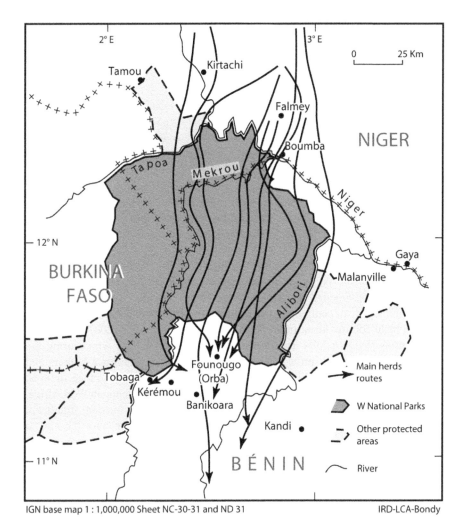

IGN base map 1 : 1,000,000 Sheet NC-30-31 and ND 31   IRD-LCA-Bondy

Figure 8.4   Routes used by pastoralists to cross the W National Parks during the severe droughts of 1973 and 1984

During the drought of 1973, no one forbade the presence of cattle in the W National Park. The herds could enter and cross it to move to the dry season pastures over there in Burkina Faso, and then come back through the Park. When rangers intercepted herds, they did not hurt the pastoralists. They simply asked them to leave and let them go. At the time, rangers were not mean.
(Survey in Kollo, Niger, October 2005).

However, the financial support of the European Union and various international organisations of nature conservation have, in recent years, led to the reinforced exclusion of pastoralists, with the increased surveillance and repression of illegal herders. Thus, in 2002, the Benin government forbade the arrival of herders from Niger, so as to combat illegal grazing in protected areas, the W National Park in particular.

> *In the past, we had good relations with the rangers of the W National Park in Benin. We used to spend whole days there. Today, when we go there, we have to hide. When you come across rangers, they arrest you. In the park in Niger, they confiscate the cattle. In Benin, they shoot on the cows. Before, when we entered the park, we entered the wide bush and we weren't scared, except of the lions. Today, we're scared of the rangers; we always think about them. If we're not careful, they arrest us today, they arrest us tomorrow. Last year, 30 youngsters from Niger were arrested in the park in Benin and imprisoned in Kandi.*
> (Survey in Falmey, Niger, September 2003).

Everywhere in West Africa, tensions run high between the Forestry Department and the pastoralists, particularly during years with severe drought, when the pastoralists are compelled to enter protected areas. This exclusion policy reflects the colonial legacy of centralised and paramilitary forestry departments unfamiliar with the local communities. In northern Cameroon, Seignobos (2001) mentioned an infamous Water Affairs and Forestry Department Director who, during the implementation of large protected areas, refused to take traditional hunters and farmers' organisations into account. It could be said that intolerance or even hostility towards the local populations is a generalised phenomenon in West Africa.

The eviction of pastoralists during the colonial era is also a significant characteristic of the history of protected areas in East Africa, with the aim to conserve wildlife. However, since only a quarter of the wildlife lives permanently inside protected areas, this means that, over time, wildlife and domestic cattle have to coexist in the same areas for part of the year at least (Bourn and Blench 1999: 34). The attention given to the large savannah of East Africa and to the Maasai led, after Independences, to the implementation of an official narrative associating nature conservation and development (Rusten Rugumayo 2000). In this regard, the French and Anglo-Saxon policies on nature conservation have always differed: the former advocating the separation and exclusion of local populations and pastoralists from nature conservation, and the latter advocating the opposite. Symbolically, this opposition reflects the principles underlying protected areas as founded in north Cameroon in the 1950s–1960s, and in Ngorongoro during the same period. The latter echoed a philosophy of symbiosis involving pastoralism and wildlife protection (Fosbrooke 1972), even if it is true that it did not lead to its effective implementation on the ground.

This opposition is also reflected in the differing administrative structures of national parks, which are more decentralised in East Africa. Thus, the area of Ngorongoro is administered by a local authority, i.e. the Ngorongoro Conservation Area Authority. Although its name refers only to nature conservation, this administration is also responsible for the economic, and particularly the pastoral, development of the area under its jurisdiction. As to the pastoralists, organised themselves into a Pastoral Council composed of community leaders and technical service representatives.

In fact, behind this façade of decentralised and participative form of management, lies a whole legacy of tensions, resentments and disillusionment between the conservationists and the pastoralists (Rusten Rugumayo 2000). The Pastoral Council does not have any real powers, and the agents of the local authority are more preoccupied with the protection of wildlife and promoting tourism than with pastoral development. *De facto*, several decisions made by this authority were perceived by the pastoralists as being hostile towards them: cattle were forbidden to enter the crater of Ngorongoro; bush fires were prohibited; and the areas between those allocated for wildlife and those left to cattle were to fall under a zoning process. Conversely, the Conservation Area Authority was hostile towards the lifting of a ban on cultivation and hindered a pastoral development project financed by an international NGO. As for the pastoralists of Ngorongoro, they suspect the local administration of always seeking to evict them from the pastures shared by both wildlife and the cattle. In East Africa, the relations between a great number of pastoralists and the decentralised administration of a protected area can be fraught with suspicion, and overshadowed by conflicts.

Generally, the pastoralists do not appreciate the many restrictions imposed on the cattle herds while wildlife remains free to move around. Several studies conducted on Ngorongoro at the end of the previous century showed a reduction in the number of Maasai cattle (Homewood and Rodgers 1991; Arhem 1985; Homewood et al. 1987). At the time, the Maasai were in the process of becoming poorer. Recently, the situation has supposedly worsened to such an extent that Homewood admitted that "Ngorongoro is in distress" (verbal information 2004). Behind the narrative concerning wildlife-cattle coexistence, the administrative logic of nature conservation has been winning over the pastoralists and putting them at a disadvantage compared to pastoralists in other regions. In fact, the sharing of natural resources between wildlife and domestic cattle has led to the dissatisfaction of both pastoralists and conservationists, with the former suffering the negative consequences of coexisting with wildlife. To slow down their impoverishment, these pastoralists began cultivating crops, thereby aggravating the tensions with the defenders of conservationist interests.

Yet, even in cases of difficult coexistence, international nature protection organisations continue to advocate associating conservation with pastoralism in East Africa (Bourn and Blench 1999: 36). In their opinion, pastoral areas shield wildlife from agricultural expansion, which is the real threat to conservation. Certain pastoral areas act as the last remaining corridors and zones for the dispersal of wildlife across national parks (Bourn and Blench 1999: 115), the reason why, in East Africa, international organisations view pastoralists as the allies of wildlife in the face of large-scale agriculture. As a result, from now on pastoralism needs to be considered not only in relation to conservation, but also to agriculture.

From Coexistence to Participation

Currently, the dominating narrative is that the future of protected areas must remain associated with the development of the populations living on the outskirts of such protected areas. If these populations are to respect wildlife in protected areas, while suffering constraints from it, it is essential that they are allocated a share of the benefits generated by tourism. For example, this profit-sharing principle is at the root of the Campfire Programme in Zimbabwe (Communal Areas Management Programme for Indigenous Resources) which provoked many publications and, more generally, the 'community-based' conservation policies subsequently multiplied in Southern Africa (Hulme and Murphree 2001).

In the same logic, the narratives and initiatives in favour of the participation of the West African local populations in the management of protected areas are multiplying. In Burkina Faso, within the framework of the 'Terroir Management' approach, village groups and committees have been established to assist the Forestry Department. These committees are called upon to combat poaching, so-called "unplanned" bush fires and the illegal entry of cattle. However, while the committee members are volunteers, recruited from the villagers, the pastoralists' representation is in the minority, or lacking altogether. Transhumant herders are never associated with these organisations and, consequently, represent a frequent target for their interventions. These lead to cattle found in the wrong place being impounded, which represents a form of repression particularly dreaded by the pastoralists but profitable for the village committees (Kiéma 2007).

Again in Burkina Faso, another form of involvement by the local populations with conservation finds expression in the creation of village hunting areas or Zones villageoises d'intérêt cynégétique (ZOVIC) on their lands. ZOVICs with the most game are leased to private companies associated with tourism or hunting. Thus, in a vast village territory adjacent to the W National Park, a ZOVIC and a private reserve were marked out, with both enclaves being connected to the park, thereby "making the access to these pastures all the more difficult" (Sawadogo 2006: 29). From year to year, herders have been reprimanded for entering new ZOVICs, ignoring their existence and, *a fortiori*, their boundaries.

In Benin, when the resident populations of forest reserves realised that the classification of the forests would dispossess them from their customary rights, they turned the situation around by imposing taxes on transhumant herders. In this context, the participative approach in the management of protected areas triggered power struggles between local actors with diverging interests (Houndagba et al. 2007: 332). In northern Cameroon, the village hunting areas located between the great national parks are entrusted to private safari companies. Their managers either expel or shoot cattle found in the hunting areas. Yet the herders, even the sedentary ones, rely heavily on the pastures of the hunting areas which they use as 'waiting rooms', before leading their cattle to the farmers' stubble fields. In this light, they often end up making 'arrangements' with the local agents of the hunting areas, as long as the safari tourist season has not begun (verbal information of Kossouma

Liba'a, June 2007). Generally, here and there in West Africa, local initiatives for the protection of wildlife are largely hostile to transhumant pastoralists.

Conversely, in East Africa, as landholders, pastoralists seem to be in a better position to make the most of situations found on the outskirts of national parks. In Kenya, the small community-based wildlife reserve formula is actually preferred as the new form of partnership with the Maasai, the objective being to make them benefit from tourism which, in this case, typically illustrates the commodification of nature. However, the example of a small reserve close to the National Park of Amboseli, also illustrates how a pastoral group relying on a private operator is actually not in an adequate position to engage in wildlife tourism, which could lead the group to suffer various trials and tribulations (Rutten 2002).

On the outskirts of another national park in Kenya, the Maasai have constituted wildlife associations that are also negotiating the concession of land to private tour operators. In this case, the income from tourism can be lucrative through site rental, tourist taxes, payment for overnight stays and employment. However, through a recent study, it was discovered that the majority of pastoralists only make a small profit from the letting of their pastures (Thomson and Homewood 2002). The pastoralists who live close to the viewing areas where operators bring their clients are paid a larger share of the dividends from tourist-generated income, and progressively less goes to the pastoralists who live further away. Moreover, a large portion of the income generated by tourism is used to cover management and personnel costs, while another share is paid to the local elites who act as the 'brokers' of the tourist operators. Inequalities within Maasai society are becoming larger, between the beneficiaries of tourist-generated income and the rest. Consequently, the latter are tempted to rent out their land to agricultural entrepreneurs, another way of bringing in a high income, even if this antagonises tourism. On the whole, traditional stock-rearing as undertaken by Maasai pastoralists, does not bring in revenues comparable with these speculative activities. While the Maasai of Kenya define themselves through pastoralism, this activity could soon be limited to a cultural role only. In time, land control and land-generated revenues will become major issues.

The very long coexistence of the pastoralist herds and wildlife in East Africa has led the authorities to promote policies associating development and conservation. However, the Maasai, as former pastoralists, have engaged in the diversification of their income by undertaking agricultural activities, initially to produce food not only for themselves, but also for sale. In order to protect their new crops from being damaged by wild herbivores, the Maasai have erected solar-powered electric fences financed by development programmes. The result of this spatial compartmentalisation has restrained wildlife movement and closed usual migration routes, sometimes causing an excess of animals in one place, as is the case for elephants in Amboseli (Western 1994).

The policy for the participation of local populations in protected areas, by generating revenues from tourist activities, has already been criticised. The premise that economic revenues should be sufficient to ensure that local populations adopt a positive attitude towards wildlife is a utilitarian preconception. Yet, the relations between the populations and the environment are not only aligned with material preoccupations, they also refer to cultural values. Moreover, the fact that local populations have gained access to the natural resource market has created social dynamics of inequality (Galvin et al. 2006). Among the pastoralists of East Africa, these dynamics have given rise to divisions within the pastoral communities. Finally, in West Africa, pastoralists are excluded from the environmental market by other groups that implement monopolising strategies, legitimated by former domination-based relations.

Conclusion: Engaging in Tourism or Maintaining Pastoralism?

Despite the fact that they practice the same activity, the situation of the pastoralists vis-à-vis protected areas continues to differ in West and East Africa. The way land is controlled plays an important role in this regard. By controlling zones on the outskirts of protected areas, local populations can take more or less advantage of tourist-generated revenues. In this new form of association between pastoralism and nature conservation, the Maasai of Kenya seem better placed, thanks to land rights protected better there than anywhere else, and also thanks to sustained tourist activity. The Maasai of Tanzania, on the other hand, are in a less favourable position, due to the national policy reform on the suppression of customary law, for which they are bearing the costs in the pastoral areas. As to the Fulani of West Africa, they are in a marginal position everywhere vis-à-vis the new conservation initiatives involving local communities. As local populations holding fundamental and usage rights on lands situated on the outskirts of protected areas, the villagers have been excluding the pastoralists, particularly those practicing transhumance, from the associations that intend to take advantage of the wealth the wildlife has to offer. Whereas the Forestry Department did not normally see differences in the populations, the current policies for local participation and power decentralisation are ratifying local logics resulting in local divisions and exclusions. In this light, the Fulani of West Africa stand no chance of deriving any dividends from the village wildlife reserves.

On the other hand, there seems to be an inverse relationship between the degree of evolution of those who maintain pastoral activities and those who take part in local protected areas. In this regard, the involvement of the Maasai of Kenya in tourism is part of a general diversification of their activities, one that goes hand in hand with a progressive loss of pastoral identity (built mainly during the 19th century and just as easily lost). Consequently, ethnic identity is more easily asserted in the political sphere than in a specific activity. The Maasai of Tanzania, on the contrary, have been agro-pastoralists for a long time. As to the Fulani of West Africa, sometimes they evolve from pastoralism to agro-pastoralism, but also vice-versa. The Fulani, who have been excluded from all current speculative activities involving the environment, have kept to pastoralism mainly because they have been relegated to this activity. And no association or integration of pastoralism with nature conservation has been envisaged to date, even if pastoralism is by definition close to nature.

Yet, the pastoralists are often in possession of cattle breeds that are equally as endangered as the wildlife. Indeed, these fairly unproductive breeds, in terms of milk or meat, are being progressively replaced by other breeds more highly-rated by the markets. The *ankole* race in East Africa, as well as the Red Bororo and the White Fulani breeds in West Africa could be marginalised. All these are typical pastoral breeds unfit for settlement, or multiple activities. By proposing to integrate the safeguard of these bovine breeds into the protection of wildlife, Bourn and Blench (1999) have proposed the notion of 'co-conservation'. The likely compatibility between the safeguard of bovine breeds and the protection of wildlife, leads to the idea of a multi-purpose yet specific protected area in the savannah environment. Indeed, the practice of pastoralism and the maintenance of biodiversity entail, at the same time, the existence of seasonal vegetation fires. Wild or domestic herbivores associated with vegetation fires have contributed to building the biodiversity of the savannah ecosystems; their coexistence should be seen as a contributing factor in the maintenance of biological diversity. Rather than fuelling the confrontation between conservation and pastoralism, policies on the management of protected areas should, perhaps, rather take into account the threat of agricultural expansion on these two forms of land use and their relations with the environment.

References

Amadou B., Boutrais J., 2005 – *Les transhumances d'éleveurs nigériens dans le parc du W. Logiques pastorales et de conservation de la nature*. Conference IRD "Aires protégées", Ouagadougou, 28–30 November 2005.

Arhem K., 1985 – *Pastoral man in the garden of Eden. The Maasai of the Ngorongoro Conservation Area, Tanzania*. Uppsala, Scandinavian Institute of African studies, p. 123.

Bassett T., 2002 – "Patrimoine et territoires de conservation dans le nord de la Côte d'Ivoire". *In* Cormier-Salem M.-C. *et al.* (eds.), *Patrimonialiser la nature tropicale. Dynamiques locales, enjeux internationaux*. Paris, IRD Éditions, coll. Colloques et séminaires: 323–342.

Bassett T. J., Turner M. D., 2007 – sudden shift or migratory drift? fulbe herd movements to the Sudano-Guinean region of west Africa. *Human Ecology*, 35: 39–49.

Benoit M., 1988 – Espaces "francs" et espaces étatisés en Afrique occidentale. *Cahiers Sciences Humaines,* 24 (4): 503–519.

Benoit M., 1999 – *Statut et usages du sol en périphérie du parc national du W au Niger. T. 4: Peuplement et genres de vie dans le Gourma oriental avant la création du parc national du W du Niger*. Paris/Niamey, IRD Éditions, p. 96.

Bernardet P., 1999 – "Peuls en mouvement, Peuls en conflit, en moyenne et haute Côte d'Ivoire, de 1950 à 1990." *In* Botte R., Boutrais J., Schmitz J. (eds.), *Figures peules.* Paris, Karthala: 407–444.

Blench R., 2004 – *Natural Resource Conflicts in North-Central Nigeria. A handbook and case studies.* Cambridge, Mallam Dendo, p. 186.

Bourn D., Blench R. (eds.), 1999 – *Can livestock and wildlife co-exist? An interdisciplinary approach.* London, ODI-ERGO.

Boutrais J., 2007 a – "Pasteurs d'Afrique de l'Ouest et de l'Est face à une catastrophe ; la peste bovine de 1890". *In* Landy F., Lézy E., Moreau S. (eds.), *Les raisons de la géographie.* Paris, Karthala: 175–192.

Boutrais J., 2007 b – Crises écologiques et mobilités pastorales. Les Peuls du Dallol Bosso. *Sécheresse*, 18 (1): 5–12.

Bouyer J., 2006 – *Écologie des glossines du Mouhoun au Burkina Faso. Intérêt pour l'épidémiologie et le contrôle des trypanosomes africaines.* PhD., University of Montpellier-II.

Bouyer J., Guerrini L., Desquesnes M., De La Rocque S., Cuisance D., 2006 – Mapping African animal trypanosomosis risk from the sky. *Veterinary research*, 37: 633–645.

Doutressoulle G., 1947 – *L'élevage en AOF.* Paris, Larose, p. 298.

Ford J., 1971 – *The role of trypanosomiases in African ecology. A study of the tsetse fly problem.* Oxford, Clarendon Press, p. 568.

Fosbrooke H., 1972 – *Ngorongoro: the eighth wonder.* London, Deutsch, 240 p.

Fournier A., Kaboré-Zoungrana C., 2007 – *Gestion des activités d'élevage et des feux de végétation et conservation de la biodiversité au Burkina Faso.* Research project Corus.

Fournier A., Millogo-Rasolodimby J., 2007 – "Une végétation menacée ou modelée par les hommes ?" *In* Fournier A., Sinsin B., Mensah G. A. (eds.), *Quelles aires protégées pour l'Afrique de l'Ouest ? Conservation de la biodiversité et développement.* Paris, IRD Éditions, coll. Colloques et séminaires, CD-ROM: 33–46.

Galvin K. A. *et al.*, 2006 – Integrated modelling and its potential for resolving conflicts between conservation and people in the rangelands of East Africa. *Human Ecology*, 34 (2): 155–183.

Homewood K. M., Rodgers W. A., 1984 – Pastoralism and conservation. *Human Ecology*, 12 (4): 431–441.

Homewood K. M., Rodgers W. A., 1991 – *Maasailand ecology, pastoralist development and wildlife conservation in Ngorongoro, Tanzania.* Cambridge, Cambridge University Press, p. 318.

Homewood K. M., Rodgers W. A., Arhem K., 1987 – Ecology of pastoralism in Ngorongoro Conservation Area. *Journal Agric. Sciences*, 108: 47–72.

Houndagba C. J., Tente A. B., Guedou R., 2007 – "Dynamique des forêts classées dans le cours moyen de l'Ouémé au Bénin: Kétou, Dogo et Ouémé-Boukou". *In* Fournier A., Sinsin B., Mensah G. A. (eds.), *Quelles aires protégées pour l'Afrique de l'Ouest ? Conservation de la biodiversité et développement*. Paris, IRD Éditions, coll. Colloques et séminaires, CD-ROM: 325–335.

Hulme D., Murphree M. W. (eds.), 2001 – *African wildlife and livelihoods. The promise and performance of community conservation*. Portsmouth/Oxford, Heinemann/James Currey, p. 336.

Kiéma S., 2007 – *Élevage extensif et conservation de la diversité biologique dans les aires protégées de l'Ouest burkinabé. Arrêt sur leur histoire, épreuves de la gestion actuelle, état et dynamique de la végétation*. Ph.D., University of Orléans, p. 657.

Kiéma S., Fournier A., 2007 – "Utilisation de trois aires protégées par l'élevage extensif dans l'ouest du Burkina Faso". *In* Fournier A., Sinsin B., Mensah G. A. (eds.), *Quelles aires protégées pour l'Afrique de l'Ouest ? Conservation de la biodiversité et développement*. Paris, IRD Éditions, coll. Colloques et séminaires, CD-ROM: 445–451.

Luxereau A., 2004 – Des animaux ni sauvages ni domestiques, les "girafes des Blancs" au Niger. *Anthropozoologica*, 39 (1): 289–300.

MacKenzie J. M., 1988 – *The empire of nature. Hunting, conservation and British imperialism*. Manchester, Manchester Univ. Press, p. 340.

Pierre C., 1906 – *L'élevage dans l'AOF*. Paris, A. Challanel.

Rusten Rugumayo C., 2000 – *The politics of conservation and development; on actors, interface and participation; the case of Ngorongoro Conservation Area, Tanzania*. Ph.D., Trondheim, p. 398.

Rutten M., 2002 – *Parcs au-delà des parcs. Écotourisme communautaire ou nouveau revers pour les pasteurs maasai au Kenya ?* London, IIED, p. 28.

van Santen J., 2008 – "Giving a Voice to the Elephants"; the intricate relation between wildlife, local populations and global actors in north Cameroon". *In* van Santen J. (ed.), *Development in place. Perspectives and challenges*. Amsterdam, Aksant: 281–309.

Sawadogo I., 2006 – *Transhumance et pratiques pastorales sur le terroir de Kotchari en périphérie du parc du W du Burkina Faso*. Master thesis, INA-PG/Ecopas.

Seignobos C., 2001 – *Les mots du développement. Histoire du développement du Nord-Cameroun*. Paris, Habilitation à diriger des recherches, University of Paris-I, p. 167.

Sinsin B., 1998 – "Transhumance et pastoralisme". *In* Sournia G. (ed.), *Les aires protégées d'Afrique francophone*. Paris, ACCT/Éditions Jean-Pierre de Monza: 26–31.

Sougnabé P., Ali Brahim B., Hassane Mahamat H., 2004 – *Étude sur les pratiques pastorales dans et autour de la forêt classée de Yamba-Berté*. Farcha, LRZV.

Sournia G., 1998 – *Les aires protégées d'Afrique francophone*. Paris, ACCT/ Éditions Jean-Pierre de Monza, p. 272.

Thomson M., Homewood K. M., 2002 – Entrepreneurs, elites and exclusion in Maasailand: trends in wildlife conservation and pastoralist development. *Human Ecology*, 30 (1): 107–138.

Toutain B., de Wisscher M.-N., Dulieu D., 2004 – Pastoralism and protected areas: lessons learned from Western Africa. *Human Dimensions of Wildlife*, 9: 287–295.

Waller R. D., 1988 – "Emutai; crisis and response in Maasailand". *In* Johnson D., Anderson D. M. (eds.), *The ecology of survival. Case studies from North-East African history*, London, Lester Cook and Boulder: 73–114.

Western D., 1982 – Amboseli National Park; enlisting landowners to conserve migratory wildlife. *Ambio*, 11 (5): 302–308.

Western D., 1994 – "Ecosystem conservation and rural development; the case of Amboseli". *In* Western D., Wright R. M. (eds.), *Natural connections. Perspectives in community-based conservation*. Washington, Island Press: 15–52.

Conclusion
Understanding Protected
Areas in Globalisation

Hervé Rakoto Ramiarantsoa and Estienne Rodary

This publication reveals a great variety of situations resulting from the experimentation of sustainable development approaches. As such, the sustainable development approach to natural resource management is a reflection of the trend towards the ecologisation of development policies (Rodary and Milian). This raises questions regarding the various influences spatial, temporal and actors' scales have on the dynamics of protected areas (Carrière et al., Méral et al.).

Naturally, the case studies report some situations that are already known. This is the case for the tensions still affecting the relationships between the administrations that manage the resources and the populations residing next to these very resources. This is also the case for specific situations: of scale interactions, as in the Amazon (Aubertin and Filoche, Albert et al.); of agents as in Africa between wildlife and cattle (Boutrais); or of specific areas such as coastal areas (Chaboud et al.). But the cases presented here also report more original findings, such as the role of community-based management in protected areas worldwide (Rodary and Milian). These studies are in line with both the general context of sustainable development, and the specific context of the history of protected areas. They question the extent to which the environment either emerged as the main issue of land policies, or remained marginal in development policies. In this context, this conclusion will develop two central points mentioned in the introduction. First, the coherence between the definitions guiding the objectives of protected areas governance and the tools that are used; second, the perspectives of radical transformation in conservation policies with regard to the politics of participation of different stakeholders.

Contextualising Protected Areas

On reading the various texts of this publication, the first conclusion is the importance of the dynamics of globalisation in biodiversity conservation. For several decades, although with a notable reinforcement after the Summit of Rio in 1992, globalisation has been both structuring and determining discourses. It has highlighted the necessity to combine "the needs of the indigenous people with those of healthy ecosystems" (Worldwatch Institute 2005). Globalisation connects the natural sites of the planet through ecological networks. Bonnin speaks of the "supranational value" granted to natural landscapes. In line with the logic of capital, the acquisition of development rights over land reinforces the contractual exclusion of different social groups. However, the studies highlight the unequal appropriation by societies of the opportunities created by globalisation[1]. The analyses also focus on some ambiguities in conservation practices: the blurry dimension of the participation, the interference between the community-based and the 'back to the barriers' movements, and the discrepancies between definitions and tools in conservation. These are but a few examples that show the need to contextualise protected areas.

Therefore, whereas the main trends of protected area management policies are driven on a global scale, specific studies actually show that their trajectories are not linear. The case studies reveal several types of interweaving: of rights (Aubertin and Filoche), powers (Chaboud et al.) and management (Albert et al.). Moreover, while goals can be clearly stated, divergences appear during their implementation that are linked to different geographic contexts. The first element relates to the heritage bequeathed to the protected areas, which differentiates them from one another in their appropriation, their constraints and their dynamics. Boutrais recalls a fundamental difference between West and East African pastoralism and park management. West Africa is imbued with a nature conservation policy of French origin that divides and excludes, whereas the Anglo-Saxon policy has already several decades of experiences in associating local communities with conservation. A similar heritage issue is to be found in the management of the 'mega-biodiversity' in Madagascar. The challenge of conservation requires a "major evolution of the mode of governance", combining the management practices inherited from the colonial administration, the more recent French co-operation, with Anglo-Saxon financing through international conservation NGOs. Another element that ought to be mentioned is when protected area management overlaps with other administrative structures of a territory. Depending on the responsibilities and involvement of the latter, we observe 'local' initiatives that bring a special touch to governance of protected areas; the analysis of the Guyana Amazonian Park (Aubertin and Filoche) illustrates this point.

However, we need to point out that, irrespective of the context, protected areas concern local as well as global realities, and their roles in conservation need to be considered at every level. First, on the spatial scale, not only because the coexistence of species (e.g. wildlife and cattle) on a large scale brings separate trajectories at the local scale (Boutrais), but also because from now on one must understand how protected areas fit into inclusive systems, such as the ecological networks (Bonnin). Second, the temporal scale, whether it concerns the funding of these areas (Méral et al. in this regard note the inadequacy between the temporalities of aid linked to the typical five-year international programmes, and

---

1   On that subject, see the special issue on capitalism and conservation in Brockington and Duffy (2010).

the longer time required for the sustainability of biodiversity conservation), the sequential utilisation of the same resources (e.g. the herds belonging to the Maasai in East African parks), or still the unequal temporalities that characterise corridors and the distinction between a conduit and a habitat (Carrière et al.). Third, and finally, is the scale of actors intervening in the field of protected areas, including the dominant status of the supranational, and the heightened importance of conservation NGOs that end up as financial intermediaries in the new structure of payments for ecosystem services (Méral et al.). The significance of the interference of these scales leads to consider at the same time the various levels operating in the delimited field of conservation policies, and what, outside that field, influences conservation institutions and governance. Bonnin helpfully reminds us that the capacity of the Natura 2000 network to realise its objectives "depends particularly on the maintenance or restoration of an appropriate matrix of territory, both in and between the sites".

Coherence Between Definitions and Tools

All studies show that sustainable development is both strong in discourses and still is rarely reflected in the governance of protected areas. When it comes to the categories of protection, the decision-making and the tools implemented, sustainable development is generally very marginal; the prioritisation of local demands is a reality only in specific contexts; the consultation process with the different stakeholders is barely balanced; and the link between information and action in regard to sustainability is far from evident.

A first point concerns the definition of what needs to be conserved. On what information is this choice based? It appears that local knowledge is given little consideration, despite the fact that it often reveals practices that enrich biodiversity. Recognising local knowledge, or even re-qualifying it as 'traditional' knowledge, poses the problem of the valorisation of the local scale by public policies. Boutrais examines this in the light of pastoralism in Africa. For the shepherds who are stakeholders in the management of the environment (e.g. fire management, now recognised by biological science as a useful disturbance for, in its absence, environments close up and loose their pyrophyte species), protected areas are resources of food and security for their animals. Moreover, the shepherds' livestock typically includes pastoral breeds that have become rare, and cannot tolerate a sedentary lifestyle. But neither the shepherds' perception of the land nor the fact that they own endangered bovine breeds, is taken into account by the protected area administrations. Reconciling nature conservation with pastoralism remains inconceivable. Whereas both domains would benefit by complementing each other, particularly when confronted with agriculture, the real threat to both. Generally, what constitutes a major local constraint is downgraded by managerial criteria decided at higher levels.

This directly concerns the second point, on the role attributed to scientific knowledge. Chaboud et al. insist there is an indispensable requirement for adequate scientific expertise in order to specify the objectives and the methods of any sustainable policy. Although such a requirement appears obvious, too often only 'standardised' academic science is employed when defining projects. The fact that local knowledge is not called upon, generates problems concerning their social acceptability. Moreover, unreliable data, and unverified results, constitute the foundations of certain practices. The text of Carrière et al. sheds light on the confusing contradictions between the credit given to the concept of corridor, now an essential part of conservation policy, and the many criticisms concerning this concept, making it "barely workable". The eviction of pastoralism from protected areas in Africa on the grounds of competition between wildlife and cattle is just as problematic; whereas a contrary perception acknowledges that wildlife plays a role in the protection of pastoral areas against the expansion of agricultural land. Some are even surprised to find, in Guyana, parks with fragmented cores, breaking the ecological rule according to which there is an exponential link between the number of species and the surface area. Many cases illustrate the misalignment of political decisions and scientific expertise, power and knowledge (see notably Fairhead and Leach 1996). In this sense, political choices in recent conservation policies (participation, ecological networks, etc.) seem to perpetuate previous use of the scientific expertise.

This raises a third point, which has to do with the relationship between global and local scale, i.e. how decisions made on a global scale impact the communities residing next to the resources affected by these decisions. The polarity is rarely reversed, and occurs only in exceptional cases (such as the international networks instituted by the Amerindian communities). Yet, the constantly evolving importance of actors at the international level is an intrinsic cause of disconnection with the local level. The dimension taken by the international scale appears across several issues. It accompanies the increasing leverage of global NGOs, in tune with the latest development towards 'large scale' approaches in conservation (Wolmer 2003; Hughes 2005; Büscher 2009). Chapin (2004) points out that although recently (1950–2000) the funds available for conservation have decreased by 50%, those allocated to Conservation International, The Nature Conservancy and the World Wide Fund for Nature have experienced a relative, as well as an absolute, increase in value. The relationships between these three organisations and private firms and bilateral/multilateral undertakings and financing, enables them to control the redistribution of resources allocated for conservation carried out by local organisations: which *de facto*, have their initiatives limited. Furthermore, these large NGOs create partnerships with international institutions outside the conservation field (e.g. USAID and the World Bank). Because the environment is a conditionality of aid, one can understand why the state is so reluctant to look for the interests of local populations. Yet, the example of the Brazilian Indigenous Lands (Albert et al.) illustrates that when dealing with an issue as delicate as that of the legitimacy of access to resources, modern national law can both recognise, and integrate, indigenous rights (Chaboud et al.). Another domain where we find a dichotomy between a global dominant position and a local marginalisation, concerns the payments for ecosystem services, a perspective which is yet set to become a reference point for conservation (Armsworth et al. 2007; Redford and Adams 2009). Méral et al. reveal how the economic logic of 'carbon projects' could result in programmes on large surface areas in which the indigenous populations have absolutely no control. Ought we to anticipate a similar initiative in the ecological networks that "compete" (Bonnin) with protected areas? This recent trend conceives sustainable development through territorial zoning (core zones, buffer zones and biological corridors) and through the fitting of these different levels. Naturalist scientific knowledge, when focused on spatial delimitations, elevates biophysical units of a certain size to the rank of a land use planning tool, by introducing the notion of 'ecological infrastructure'. The natural resource is no longer given consideration at the local level. Irrespective of its scientific validity, this vision alters the level of understanding of the resource, therefore changing its status by removing it from the sphere of activity of the communities. Such a detachment of initiatives from local realities reflects a much more general attitude concerning sustainable development carried out within protected areas: an approach which serves politics exclusively.

Does this signify an improvement in the way conservation policies face ecological challenges? It is difficult to offer a clear answer to this question, all the more since it raises the problem of separation of knowledge and action. The texts in this publication touch on three points. The first concerns the lack of clarity of the concepts used to define conservation tools. The most striking example is that of the controversy surrounding corridors, yet they "have been given a role to play in the conservation of forest ecosystems, particularly by overcoming the potential effects of their fragmentation, the resulting isolation of their animal and plant populations, and even their extinction". While highlighting the absence of a clear scientific assertion that corridors maintain functional connectivity, the contribution of Carrière et al. sheds light on the exploitation of the concept to obtain finance for the protection of the Malagasy forests. The second point tackles the issue of the contradictory results that science can supply, which calls for a refinement of these studies in order to identify adequate forms of management. The fragmentation of the ecological systems in Sudanese Africa is a case in point. The contact between protected areas and the open spaces that border them, allows the population of herbaceous species to be maintained on the outskirts, protected areas therefore taking on the role of vegetation reservoir. At the same time, these bordering open spaces are ecologically dangerous, with their high concentration of glossinas in a fragmented landscape with a high natural gradient. This contact area is attractive due to its wealth of available pastures, but is also dangerous because of the presence of tsetse flies. How then should it be managed according to its ecological functionality? It appears in any case that the problem is causing the argument between conservation and pastoralism to shift from the protected areas to their outskirts. Finally, the third point notes the perception, by indigenous people, of the ecological issues associated with conservation. Whereas trees and wildlife are at the centre of the foresters' interventions in protected areas, for the cattle farmer there is no incompatibility between grazing the herbaceous cover, and protecting ligneous species, because grazing leads to a reduction in bush fires. Yet the relevance of this logic goes unrecognised. These three points also highlight the difficulty in distinguishing the ecological dynamics of conservation, so as to properly appreciate them, since the intervention of man is so pervasive. In his comparison between West and East Africa, Boutrais reports how technology and social dynamics have enabled cattle farming to develop in savannah areas that were not previously frequented because of the presence of the tsetse fly. But he also reports how the fallow of farmland generates scrub spreading, which is the cause of the dramatic return of the fly and therefore bovine trypanosomosis. A few years ago, on the theme of the ecological challenges of conservation, Jacques Weber (2000) was already asking an important question relating to the balance between objective and method: "Can the social be managed biologically?" These complex combinations probably explain why, more generally as far as protected areas are concerned, "an evaluation process [...] is yet to be defined" (Bonnin).

Finally, the lack of coherence between tools is also obvious in the domain of protected area financing. The recent expansion of protected areas makes the creation of new forms of sustainable financing indispensable; the traditional resources being inadequate, particularly in the context of privatisation of the state. The future evolution of spatial governance, although not yet clearly defined, will depend on these new forms of sustainable financing. An analysis of this issue shows that the multiplication of the protected areas is a headlong rush that constitutes an incentive to search for alternative financial mechanisms. And even if the mechanism can be reversed, with the finance bringing more protected areas, in either case the tool mobilised is not finalised and is refined while used. With the Malagasy example, Méral et al. highlight the effect of incoherence between definitions and tools in relation to sustainable development: the policies and management of protected areas are decided outside the country. Looking at the autonomy of national actors, did the innovation that sustainable development policies were supposed to bring go from bad to worse?

Participation is Dead, Long Live Participation!

From the point of view of sustainable development, participation brings a social dimension to the economic objectives, and to the conservation of natural resources. Participation is thus a requirement for sustainability. Integrating it into the management of protected areas constitutes a test for sustainable development, particularly through programmes based on devolution of natural resources management. Can the participation principle change the administration of protected areas, by bringing real innovation? The answers, never straightforward, can be found at the limits of the participative approach.

The analysis of Rodary and Milian sheds light on the status attributed to community-based management, when considered over the long term. The evolution of protected areas highlights that community-based areas have experienced their highest rates of expansion when the increase of all kind of protected areas was slowing down in the 1990s. This finding could mean a rupture in the historical importance of the preservationist and fortress approaches. Even then, we would need to know whether the practices mobilised in the areas thus classified actually meet sustainable development criteria. The analysis draws up two other findings that do not follow the direction of this change. The first one concerns the number of sites. For over 40 years, with the exception of Category IV (Habitat and Species Management Area), all the forms historically registered by the IUCN experience comparable rates of expansion. The authors highlight that this even applies to Category Ia (Strict Nature Reserve), the most preservationist of all the categories. Does that mean that community-based policies have been hampered by the legacy of protected areas? The second finding recognises the limited role of Category VI (Protected area with sustainable use of natural resources), yet it was intended to promote an integrative approach between conservation and development. Can we speak of protected areas as a means of sustainable development as long as this category remains marginal?

It is in this context of a dominant participation discourse that comes along with the perpetuation of more protectionist protected areas that practices integrating the co-management of local communities with the management of resources are actually carried out. This process represents an opportunity for governments to find, on site, 'responsible' stakeholders able to deal with commitment to the institutionalised conservation of natural resources (Blanc-Pamard and Boutrais 2002). In addition to this, the political and economic benefits linked to the establishment of protected areas (Chaboud et al.), are increased by community participation, as a new norm in international policies. Thus, in various forms, the participation policies become a means to reassert the role of the state that was previously weakened from its traditional functions, just as the participation movement was explicitly aimed at avoiding the state. Therefore, participation has, paradoxically, enabled a return of the state, i.e. a concerted public policy in decentralised areas.

Moreover, involvement in participation institutions can reflect the initiatives taken by a local community to reinforce their territorial base. This is particularly evident in the Brazilian Amazon. The new image of indigenous people campaigning at the local level for an environmental cause has been used, by national and international levels in order to claim legitimacy over nature protection. Yet Albert et al. also shed light on the logics to the Amerindian groups within the framework of the peculiar protected areas that are the Indigenous Lands. While maintaining a structure responsible for the sustainable exploitation in these spaces, the local populations will use participation to become involved in institutions outside of their usual sphere, hence widening and strengthen their autonomy. These processes of re-appropriation are evident elsewhere in the world, particularly in Australia and Southern Africa, where land claims enable formerly marginalised communities to become involved in conservation as partners, and no longer as mere participants within structures that are developed and managed elsewhere (Reid et al. 2004).

As such, participation opens – albeit partially – a space for negotiation on the local scale that did not automatically exist before, and that actors knew how to appropriate, even if it presupposed the assimilation and translation of external norms and practices (Rodary 2009). Thus a second result of participation is indeed found in local political re-appropriation, offering a glimpse of the transformation of current conservation governance.

Participation does not always bring about local capacity building. When it is part of a protection programme that excludes communities in the decision process and includes them in the management, participation contributes to inequalities between stakeholders. Such standardisation pays little heed to the priorities at local level, despite the fact that participation is actually supposed to prioritise the capacities of local communities. This form of participation is ineffective considering that "the needs of the indigenous people [...] must be more efficiently integrated into the conservation programmes" (Collective 2005). This is illustrated by several aspects of the analysis of the Guyana Amazonian Park, even though they are actually presented as 'participative democracy' (Aubertin and Filoche). The fact that the Wayana's goal to see their villages included within the core area of the park was refused is an example of this.

Even though the concept of participation is portrayed as an innovation in sustainable management of protected areas, it remains a truly ambiguous notion. Instead of significant progress towards the empowerment of local communities, the concept has been generating situations where participation is one of many tools in the service of policies that undoubtedly aim at conserving natural resources, but which are conceived without giving priority to local needs (see Shackleton et al. 2010a; 2010b). The result is that, among the different groups that are using these resources, the most vulnerable individuals and communities do not benefit, which goes against "poverty reduction and best distribution of wealth", which are yet at the core of sustainable development theory (Chaboud 2007). At the same time, as previously noted, participation has reintroduced political issues to the local management of natural resources, both through the state and through the empowerment of local populations. Admittedly, these populations are not homogeneous: the idea of a political process enabling the equal participation of all stakeholder, is a utopian illusion capable of mobilising people into action, but obviously impractical. Nonetheless, the fact remains that the re-appropriation of the issue of nature management by certain local actors is a notable change compared to former, often authoritarian, centralised, state policies.

The result is a dual conclusion: on the one hand, participation has revealed its limitations in meeting the expectations brought with its introduction to conservation; on the other hand, the fact that these limitations are understood and taken into account, gives participation the status of a tool amongst other tools. And if explicitly intertwined with international networks, participation can indeed make it possible to reconfigure the balance of power in favour of historically marginalised actors. It brings new institutional spaces and innovative practices valorising local actors. In this sense, the post-participations period the world of conservation is currently experiencing, might indicate the beginning of a real rupture with previous policies.

References

Armsworth P.R. et al., 2007 – Ecosystem-service science and the way forward for conservation. *Conservation Biology*, 21 (6): 1383–1384.

Blanc-Pamard C., Boutrais J., 2002 – Les temps de l'environnement. D'un sauvetage technique à une gestion locale en Afrique et à Madagascar. *Historiens et Géographes*, 381: 389–401.

Brockington D., Duffy R. (eds.), 2010 – Capitalism and Conservation. Antipode, Special issue, 42 (3): 469–799.

Büscher B.E., 2010 – Seeking 'telos' in the 'transfrontier'? Neoliberalism and the transcending of community conservation in Southern Africa. *Environment and Planning A*, 42 (3): 644–660.

Chaboud C., 2007 – Gérer et valoriser les ressources marines pour lutter contre la pauvreté. *Études Rurales*, 178: 197–212.

Chapin M., 2004 – A Challenge to conservationists. *WorldWatch Magazine*, November–December: 17–31.

Collective, 2005 – A challenge to conservationists: Phase II. From reader. *Worldwatch Magazine*, January–February: 5–20.

Fairhead J., Leach M., 1996 – *Misreading the African landscape. Society and ecology in a forest-savanna mosaic.* Cambridge/New York, Cambridge Univ. Press, p. 354.

Hughes D.M., 2005 – Third nature: making space and time in the Great Limpopo Conservation Area. *Cultural Anthropology*, 20 (2): 157–184.

Marty P., Vivien F.-D., Lepart J., Larrère R. (eds.), 2005 – *Les biodiversités. Objets, théories, pratiques.* Paris, CNRS Éditions, p. 261.

Redford K.H., Adams W.M., 2009 – Payment for ecosystem services and the challenge of saving nature. *Conservation Biology*, 23 (4): 785–787.

Reid H., Fig D., Magome H., et Leader-Williams N., 2004 – Co-management of contractual national parks in South Africa: lessons from Australia. *Conservation & Society*, 2 (2): 377–409.

Rodary E., 2009 – Mobilizing for nature in southern African community-based conservation policies, or the death of the local. *Biodiversity and Conservation*, 18 (10): 2585–2600.

Shackleton C. M. Willis T. J., Brown K., Polunin N. V. C. (eds.), 2010a – Community-based natural resource management (CBNRM): designing the next generation (Part 1). *Environmental Conservation, Special issue*, 37 (1): 1–106.

Shackleton C. M. Willis T. J., Brown K., Polunin N. V. C. (eds.), 2010b – Community-based natural resource management (CBNRM): designing the next generation (Part 2). *Environmental Conservation, Special issue*, 37 (3): 22 –372.

Weber J., 2000 – "Pour une gestion sociale des ressources naturelles". *In* Compagnon D., Constantin F. (eds.), *Administrer l'environnement en Afrique. Gestion communautaire, conservation et développement durable*. Paris/ Nairobi, Karthala/IFRA: 79–105.

Wolmer W., 2006 – "Big conservation: the politics of ecoregional science in southern Africa, Culture, Nature, Future?" Perspectives on Science and Development *In* Africa. *University of Edinburgh, April 12–13 2006*, p. 14.

Index